# *Linux for Embedded and Real-time Applications*

# *Linux for Embedded and Real-time Applications*

FOURTH EDITION

Doug Abbott

*Intellimetrix Linux for the Real World, Silver City, USA*

Newnes is an imprint of Elsevier
The Boulevard, Langford Lane, Kidlington, Oxford OX5 1GB, United Kingdom
50 Hampshire Street, 5th Floor, Cambridge, MA 02139, United States

**British Library Cataloguing-in-Publication Data**
A catalogue record for this book is available from the British Library

**Library of Congress Cataloging-in-Publication Data**
A catalog record for this book is available from the Library of Congress

ISBN: 978-0-12-811277-9

For Information on all Newnes publications
visit our website at https://www.elsevier.com/books-and-journals

ELSEVIER
Book Aid
International
Working together
to grow libraries in
developing countries
www.elsevier.com • www.bookaid.org

*Publisher:* Mara Conner
*Acquisition Editor:* Tim Pitts
*Editorial Project Manager:* Naomi Robertson
*Production Project Manager:* Kiruthika Govindaraju
*Cover Designer:* Mark Rogers

Typeset by MPS Limited, Chennai, India

# Dedication

On a personal level, this book is dedicated to the two most important people in my life:

> **To Susan**, my best friend, my soul mate. Thanks for sharing life's journey with me.
>
> **To Brian**, theatrical lighting designer, actor and musician, future pilot, and all-around great kid. Thanks for keeping me young at heart.

> *Never doubt that a small band of committed citizens can change the world. Indeed, it is the only thing that ever has.*
>
> Margaret Mead

On a professional level, the book is dedicated to those who made it possible; open source programmers everywhere, especially that small band of open source pioneers whose radical notion of free software for the good of the community, and just for the fun of it, has indeed changed the world.

# Contents

# Preface

> Microsoft sells you Windows Linux gives you the whole house
>
> ***Unknown***

Much has happened in the Linux world since the third edition of this book was published in 2013, which of course is the motivation for a new edition. The kernel continues to evolve and improve. Open source hardware is proliferating with the likes of Arduino, BeagleBoard, and Raspberry PI, among others. Projects that aim to ease the embedded development process and improve productivity, such as Linaro and Yocto, have reached a level of maturity that makes them genuinely useful.

Linux still hasn't made much of a dent in the world of desktop computing, but it has achieved notable success in the embedded space where I work. A VDC Research study from March 2015 suggests that "open source, freely and/or publicly available" Linux will grow from a 56.2% share of embedded unit shipments in 2012 to 64.7% in 2017. The "Internet of Things" phenomenon is helping drive that growth.

I began the preface to the first edition by confessing that I've never really liked Unix, considering it deliberately obscure and difficult to use. Initially, Linux did little to change that impression, and I still have something of a love/hate relationship with the thing.

Linux is indeed complex and, unless you are already a Unix guru, the learning curve is quite steep. The information is out there on the web, but it is often neither easy to find nor readable. There are probably hundreds of books in print on Linux, covering every aspect from beginners' guides to the internal workings of the kernel. By now there are even a number of books discussing Linux in the embedded and real-time space.

I decided to climb the Linux learning curve partly because I saw it as an emerging market opportunity, and partly because I was intrigued by the Open Source development model. The idea of programmers all over the world contributing to the development of a highly sophisticated operating system just for the fun of it is truly mind-boggling. Having the complete source code not only allows you to modify it to your heart's content, it allows you (in principle at least) to understand how the code works.

Open Source has the potential to be a major paradigm shift in how our society conducts business, because it demonstrates that cooperation can be as useful in developing solutions to problems as competition. And while I doubt that anyone is getting fabulously wealthy off of Linux, a number of companies have demonstrated that it is possible to build a viable business model around it.

## Audience and Prerequisites

My motivation for writing this book was to create the kind of book I wished I had had when I started out with Linux. With that in mind, the book is directed at two different audiences:

- The primary audience is embedded programmers who need an introduction to Linux in the embedded space. This is where I came from, and how I got into Linux, so it seems like a reasonable way to structure the book.
- The other audience is Linux programmers who need an introduction to the concepts of embedded and real-time programming.

Consequently, each group will likely see some material that is review, although it may be presented with a fresh perspective.

This book is not intended as a beginners' guide. I assume that you have successfully installed a Linux system and have at least played around with it some. You know how to log in, you have experimented with some of the command utilities, and have probably fired up a GUI desktop. Nonetheless, Chapter 2, Installing Linux, takes you through the installation process, and Chapter 3, Introducing Linux is a cursory introduction to some of the features and characteristics of Linux that are of interest to embedded and real-time programmers.

The book is loosely divided into three parts: Part I is largely introductory and sets the stage for Part II, which discusses application programming in a cross-development environment. Part III takes a more detailed look at some of the components and tools introduced earlier.

It goes without saying that you cannot learn to program by reading a book. You have to do it. That is why this book is designed as a practical hands-on guide. You will be installing a number of open source software packages on your Linux workstation, and there is sample code available at the book's website, https://www.elsevier.com/books-and-journals/book-companion/9780128112779.

Embedded programming generally implies a target machine that is separate and distinct from the workstation development machine. The principal target environment we will be working with in this book is the BeagleBone Black, an ARM-based single board computer (SBC) introduced in Chapter 6, Eclipse integrated development environment. This is an open source, widely available, relatively inexpensive device with capabilities typical of a real-world embedded product.

## What's New in this Edition

- The BeagleBone Black replaces the Mini2440 as the target single board computer.
- All of the software is downloaded from primary sources, no more DVD.
- We take a closer look at two approaches to making the Linux kernel real-time.

## Personal Biases

Like most computer users, for better or worse, I've spent years in front of a Windows screen and, truth be told, I still use Windows for my day-to-day computing. But before that I was perfectly at home with DOS, and even before that I hacked away at RT-11, RSX-11, and VMS. So it's not like I do not understand command line programming. In fact, back in the pre-Windows 95 era, it was probably a couple of years before I finally added `WIN` to my `AUTOEXEC.BAT` file.

Hardcore Unix programmers, on the other hand, think GUIs are for wimps. They proudly do everything from the command line. Say what you will, but I like GUIs. Yes, the command line still has its place, particularly for shell scripts and makefiles, but for moving around the file hierarchy and doing simple file operations like move, copy, delete, rename, etc., drag-and-drop beats the miserably obscure Unix shell commands hands down. I also refuse to touch text-based editors like `vi` and `emacs`, although recently I've come to tolerate `vim`. Sure they are powerful if you can remember all those obscure commands. Give me a WYSIWYG editor any day.

My favorite GUI is the KDE desktop environment. It has all the necessary bells and whistles, including a very nice syntax coloring editor, not to mention a complete suite of office and personal productivity tools. KDE, an acronym for "K Desktop Environment" by the way, is included in most commercial Linux distributions. Clearly, you are free to use whatever environment you are most comfortable with to work the book's examples. But if you are new to Linux, I would recommend KDE and that is the environment I'll be demonstrating as we proceed through the book.

## Organization

The book is organized around three major sections.

*Part 1* is Chapters 1, The embedded and real-time space through 6. It is an introduction and getting started guide. You will install Linux if you do not already have it. There is an introduction to Linux itself, for those not familiar with Unix/Linux systems. You will set up and connect the target board that will be used in Part 2 to investigate cross-platform application development. Finally, you will install and get familiar with various cross-development tools including Eclipse, the open source integrated development environment.

*Part 2*, Chapter 7, Accessing hardware through Chapter 12, Configuring and building the Linux kernel, explores application programming in a cross-platform environment. We will look at accessing hardware from user space, debugging techniques, and high level simulation for preliminary debugging. We will introduce Posix threads, network programming, and Android, and we will briefly look at kernel level programming and device drivers.

*Part 3* goes on to look at various components and tools in the embedded Linux programmer's toolkit. We will look at the u-boot boot loader, BusyBox, Buildroot, and OpenEmbedded, among others. Finally, we will wrap up with an exploration of real-time enhancements to Linux.

OK, let's get on with it. Join me for a thrill-packed, sometimes bumpy, but ultimately fun and rewarding, roller coaster ride through the exciting world of embedded Linux.

PART I

# Introduction

CHAPTER 1

# The embedded and real-time space

Chapter Outline



*If you want to travel around the world and be invited to speak at a lot of different places, just write a Unix operating system.*

***Linus Torvalds***

## What is Embedded?

When I wrote the first edition of this book back in 2002, the notion of embedded computing devices was not widely appreciated. Back then many homes had at least a VCR and/or a DVD player and maybe a Tivo, all of which have at least one embedded processor. Most other appliances of the era were still basically electromechanical.

These days, virtually every appliance has an embedded computer in it. People are much more used to digital user interfaces, and while they may still not fully appreciate the nature of embedded computing, they are at least vaguely aware that they are surrounded by computers 24/7.

When I have to explain what I do for a living, I usually start by saying that an embedded system is a device that has a computer inside it, but the user of the device doesn't necessarily know, or care, that the computer is there. It is hidden. The example I usually

Linux for Embedded and Real-time Applications.
DOI: http://dx.doi.org/10.1016/B978-0-12-811277-9.00001-8

give is the engine control computer in your car. You don't drive the car any differently because the engine happens to be controlled by a computer. Oh, and there's a computer that controls the antilock brakes, another to decide when to deploy the airbags, and any number of additional computers that keep you entertained and informed as you sit in the morning's bumper-to-bumper traffic.

I go on to point out that there are a lot more embedded computers out in the world than there are PCs. In fact, recent market data shows that PCs account for only about 2% of the microprocessor chips sold every year. The average house contains at least a couple dozen computers, even if it does not have a PC.

Is an Android smartphone an embedded system? It is small, self-contained, and has limited input/output capabilities. Nevertheless, you can personalize it and download "apps" to your heart's content. In that sense, it's really more of a general purpose computing appliance. So I would say no, an Android phone is not an embedded system.

From the viewpoint of programming, embedded systems show a number of significant differences from conventional "desktop" applications. For example, most desktop applications deal with a fairly predictable set of I/O devices—a disk, graphic display, a keyboard, mouse, sound card, and network interface. And these devices are generally well supported by the operating system. The application programmer doesn't need to pay much attention to them.

Embedded systems, on the other hand, often incorporate a much wider variety of input/output (I/O) devices than typical desktop computers. A typical system may include user I/O in the form of switches, pushbuttons, and various types of displays often augmented with touchscreens. It may have one or more communication channels, either asynchronous serial, USB, and/or network ports. It may implement data acquisition and control in the form of analog-to-digital (A/D) and digital-to-analog (D/A) converters. These devices seldom have the kind of operating system support that application programmers are accustomed to. Therefore, the embedded systems programmer often has to deal directly with the hardware.

Embedded devices are often severely resource-constrained. Whereas a typical PC now has eight or more GB of RAM, and maybe a terabyte of disk, embedded devices often get by with a few MB of RAM and nonvolatile storage. This too requires creativity on the part of the programmer.

## *What is Real-Time?*

Real-time is even harder to explain. The basic idea behind real-time is that we expect the computer to respond to its environment *in time*. But what does "in time" mean? Many people assume that real-time means really fast. Not true. Real-time simply means *fast enough* in the context in which the system is operating. If we're talking about the computer that runs your car's engine, that's fast! That guy has to make decisions—about fuel flow, spark timing—every time the engine makes a revolution.

On the other hand, consider a chemical refinery controlled by one or more computers. The computer system is responsible for controlling the process and detecting potentially destructive malfunctions. But chemical processes have a time constant in the range of seconds-to-minutes at the very least. So we would assume that the computer system should be able to respond to any malfunction in sufficient time to avoid a catastrophe.

But suppose the computer were in the midst of printing an extensive report about last week's production, or running payroll when the malfunction occurred. How soon would it be able to respond to the potential emergency?

The essence of real-time computing is not only that the computer responds to its environment fast enough, but that it responds *reliably* fast enough. The engine control computer must be able to adjust fuel flow and spark timing every time the engine turns over. If it's late, the engine doesn't perform right. The controller of a chemical plant must be able to detect and respond to abnormal conditions in sufficient time to avoid a catastrophe. If it doesn't, it has failed.

I think this quote says it best:

> A real-time system is one in which the correctness of the computations not only depends upon the logical correctness of the computation, but also upon the time at which the result is produced. If the timing constraints of the system are not met, system failure is said to have occurred.
>
> ***Donald Gillies in the Real-time Computing FAQ***

So, the art of real-time programming is designing systems that reliably meet timing constraints in the midst of random asynchronous events. Not surprisingly, this is easier said than done and there is an extensive body of literature and development work devoted to the theory of real-time systems.

## How and Why Does Linux Fit in?

Linux developed as a general-purpose operating system in the model of Unix, whose basic architecture it emulates. No one would suggest that Unix is suitable as an embedded or real-time operating system. It's big, it's a resource hog, and its scheduler is based on "fairness" rather than priority. In short, it's the exact antithesis of an embedded operating system.

But Linux has several things going for it that earlier versions of Unix lack. It's "free,"[1] and you get the source code. There is a large and enthusiastic community of Linux developers and users. There's a good chance that someone else either is working or has worked on the same problem you're facing. It's all out there on the web. The trick is finding it.

## Open Source

Linux has been developed under the philosophy of Open Source software, which grew out of the Free Software Movement pioneered by Richard Stallman who founded the Free Software Foundation (FSF) in 1985. Then in 1997, Eric Raymond published a paper, and later a book, titled *The Cathedral and the Bazaar*[2], in which he argued that there are two models of software development: the cathedral—essentially top-down, and the bazaar—bottom up. In the cathedral model, source code is released, if at all, only when the software itself is released. Code developed between releases is restricted to an exclusive group of software developers, the "master craftsmen" if you will. The bazaar model has code developed over the Internet in full view of the public. As Raymond puts it, "given enough eyeballs, all bugs are shallow." That is, the more widely available the source code is for public testing, scrutiny, and experimentation, the more rapidly all forms of bugs will be discovered.

Quite simply, Open Source is based on the notion that software should be freely available: to use, to modify, to copy. The idea has been around for some 20 years in the technical culture that built the Internet and the World Wide Web, and in recent years has spread to the commercial world.

There are a number of misconceptions about the nature of Open Source software. Perhaps the best way to explain what it is, is to start by talking about what it is not.

- Open Source is not shareware. A precondition for the use of shareware is that you pay the copyright holder a fee. Shareware is often distributed in a free form that is either time- or feature-limited. To get the full package, you have to pay. By contrast, Open Source code is freely available, and there is no obligation to pay for it.
- Open Source is not Public Domain. Public domain code, by definition, is not copyrighted. Open Source code is copyrighted by its author, who has released it under the terms of an Open Source software license. The copyright owner thus gives you the right to use the code provided you adhere to the terms of the license. But if you don't comply with the terms of the license, the copyright owner can demand that you stop using the code.

[1] Linus Torvalds once said "Software is like sex. It's better when it's free." Guys, we all know sex is never free and neither is software. One way or another, you're going to pay for it.

[2] http://www.catb.org/~esr/writings/cathedral-bazaar/

- Open Source is not necessarily free of charge. Having said that there's no obligation to pay for Open Source software doesn't preclude you from charging a fee to package and distribute it. A number of companies are in the specific business of selling packaged "distributions" of Linux.

Why would you pay someone for something you can get for free? Presumably because everything is in one place and you can get some support from the vendor. Of course, the quality of support greatly depends on the vendor.

So "free" refers to freedom to use the code and not necessarily zero cost. Think "free speech," not "free beer."

Open Source code is:

- Subject to the terms of an Open Source license, in many cases the GNU Public License (see below).
- Subject to critical peer review. As an Open Source programmer, your code is out there for everyone to see, and the Open Source community tends to be a very critical group. Open Source code is subject to extensive testing and peer review. It's a Darwinian process in which only the best code survives. "Best" of course is a subjective term. It may be the best *technical* solution, but it may also be completely unreadable.
- Highly subversive. The Open Source movement subverts the dominant paradigm, which says that intellectual property such as software must be jealously guarded so you can make a lot of money out of it. In contrast, the Open Source philosophy is that software should be freely available to everyone for the maximum benefit of society. And if you can make some money out of it, that's great, but it's not the primary motivation. Richard Stallman, founder of the FSF, is particularly vocal in advocating that software should not have owners (see Appendix B).

Many leading vendors of Open Source software give their programmers and engineers company time to contribute to the Open Source community. And it's not just charity, it's good business! Even Microsoft is jumping on the Open Source bandwagon, if perhaps reluctantly. Among other initiatives, the company hosts a number of projects at github.com, although I suspect most of these are designed to work with their commercial, proprietary products.

### *Portable and Scalable*

Linux was originally developed for the Intel x86 family of processors, and most of the ongoing kernel development work continues to be on x86s. Nevertheless, the design of the Linux kernel makes a clear distinction between processor-dependent code, which must be modified for each different architecture, and code that can be ported to a new processor simply by recompiling it. Consequently, Linux has been ported to a wide range of 32-bit, and more recently 64-bit, processor architectures including:

- Motorola 68k and its many variants
- Alpha
- Power PC
- ARM
- Sparc
- MIPS

to name a few of the more popular. So whatever 32-bit, or even 64-bit, architecture you're considering for your embedded project, chances are there's a Linux port available for it, and a community of developers supporting it.

A typical desktop Linux installation runs into 10 to 20 gigabytes of disk space, and requires a gigabyte of RAM to execute decently. By contrast, embedded targets are often limited to 64 Mbytes or less of RAM, and perhaps 256 Mbytes of flash ROM for storage. Fortunately, Linux is highly modular. Much of that 10 gigabytes represents documentation, desktop utilities, and options like games that simply aren't necessary in an embedded target. It is not difficult to produce a fully functional, if limited, Linux system occupying no more than 2 Mbytes of flash memory.

The kernel itself is highly configurable, and includes reasonably user-friendly tools that allow you to remove kernel functionality not required in your application.

## Where is Linux Embedded?

Just about everywhere. A 2015 report from VDC Research claimed that in 2012 Linux powered 56.2% of embedded device unit shipments worldwide. That is expected to grow to 64.7% by 2017. Products incorporating Linux range from cell phones, personal digital assistants, and other handheld devices through to routers and gateways, thin clients, multimedia devices, and TV set-top boxes, to robots and even ruggedized VME chassis suitable for military command and control applications.

One of the first, and perhaps best known, home entertainment devices to embed Linux is the TiVo Personal Video Recorder that created a revolution in television viewing when it was first introduced in 2000. The TiVo is based on a Power PC processor and runs a "home grown" embedded Linux port that uses a graphics rendering chip for generating video.

Half the fun of having a device that runs Linux is making it do something more, or different, than the original manufacturer intended. There are a number of websites and books devoted to hacking the TiVo. Increasing the storage capacity is perhaps the most obvious hack. Other popular hacks include displaying weather, sports scores, or stock quotes, and setting up a web server.

Applications for embedded Linux are not limited to consumer products. It is found in point of sale terminals, video surveillance systems, robots, and even in outer space. NASA's Goddard Space Flight Center developed a version of Linux called FlightLinux to address the unique problems of spacecraft onboard computers. On the International Space Station, Linux-based devices control the rendezvous and docking operations for unmanned servicing spacecraft called Automatic Transfer Vehicles.

Historically, telecommunications carriers and service providers have relied on specialized, proprietary platforms to meet the availability, reliability, performance, and service response time requirements of telecommunication networks. Today, carriers and service providers are embracing "open architecture" and moving to COTS (commercial off-the-shelf) hardware and software in an effort to drive down costs while still maintaining carrier class performance.

Linux plays a major role in the move to open, standards-based network infrastructure. In 2002, the Open Source Development Lab (OSDL) set up a working group to define "Carrier Grade Linux" (CGL) in an effort to meet the higher availability, serviceability, and scalability requirements of the telecom industry. The objective of CGL is to achieve a level of reliability known as "five nines," meaning the system is operational 99.999% of the time. That translates into no more than about 5 minutes of downtime in a year.

## Open Source Licensing

Most End User License Agreements for software are specifically designed to restrict what you are allowed to do with the software covered by the license. Typical restrictions prevent you from making copies or otherwise redistributing it. You are often admonished not to attempt to "reverse-engineer" the software.

By contrast, an Open Source license is intended to guarantee your rights to use, modify, and copy the subject software as much as you like. Along with the rights comes an obligation. If you modify and subsequently distribute software covered by an Open Source license, you are obligated to make available the modified source code under the same terms. The changes become a "derivative work," which is also subject to the terms of the license. This allows other users to understand the software better, and to make further changes if they wish.

Open Source licenses are called "copyleft" licenses, a play on the word copyright intended to convey the idea of using copyright law as a way of enhancing access to intellectual property like software, rather than restricting it. Whereas copyright is normally used by an author to prevent others from reproducing, adapting, or distributing a work, copyleft explicitly allows such adaption and redistribution provided the resulting work is released under the same license terms. Copyleft thus allows you to benefit from the work of others, but any modifications you make must be released under similar terms.

Arguably, the best-known, and most widely used, Open Source license is the Gnu general public license (GPL) first released by the FSF in 1989. The Linux kernel is licensed under the GPL. But the GPL has a problem that makes it unworkable in many commercial situations. Software that makes use of, or relies upon, other software released under the GPL, even just *linking* to a library, is considered a "derivative work," and is therefore subject to the terms of the GPL and must be made available in source code form.

To get around this, and thus promote the development of Open Source libraries, the FSF came up with the "Library GPL." The distinction is that a program linked to a library covered by the LGPL is not considered a derivative work, and so there is no requirement to distribute the source, although you must still make the source available to the library itself.

Subsequently, the LGPL became known as the "Lesser GPL," because it offers less freedom to the user. So while the LGPL makes it possible to develop proprietary products using Open Source software, the FSF encourages developers to place their libraries under the GPL in the interest of maximizing openness.

At the other end of the scale is the Berkeley Software Distribution (BSD) license, which predates the GPL by some 12 years. It "suggests," but does not require, that source code modifications be returned to the developer community, and it specifically allows derived products to use other licenses, including proprietary ones.

Other licenses—and there are quite a few—fall somewhere between these two poles. The Mozilla Public License e.g., developed in 1998 when Netscape made its browser open-source, contains more requirements for derivative works than the BSD license, but fewer than the GPL or LGPL. The Open Source Initiative (OSI), a nonprofit group that certifies licenses meeting its definition of Open Source, lists 89 certified licenses on its website as of October 2016.

Most software released under the GPL, including the Linux kernel, is covered by version 2 of the license, which was released in 1991, coincidentally the same year that Linux was born. FSF released version 3 of the GPL in June of 2007. One of the motivations for version 3 was to address the problem of "tivoization," a term coined by Richard Stallman. It turns out that Tivo will only run code with an authorized digital signature. So, even though Tivo makes the source code available in compliance with the GPL, modifications to that code won't run.

Stallman considers this circumventing the spirit of the GPL. Other developers, including Linus Torvalds, see digital signatures as a useful security tool, and wouldn't want to ban them outright. The debate continues. In any case, the kernel itself is unlikely to move up to version 3 any time soon, as it would require the agreement of literally thousands of developers.

### Legal Issues

Considerable FUD[3] has been generated about the legal implications of Open Source, particularly in light of SCO's claims that the Linux kernel is "contaminated" with its proprietary Unix code. The SCO Group, formerly known as Santa Cruz Operations, acquired the rights to the Unix System V source code from Novell in 1996, although there is some dispute as to exactly what SCO bought from Novell. In any case, SCO asserted that IBM introduced pieces of SCO's copyrighted, proprietary Unix code into the Linux kernel, and is demanding license fees from Linux users as a consequence.

Ultimately, SCO's case collapsed and the company filed for Chapter 11 bankruptcy in 2007. But all of a sudden there is serious money to be made by fighting over Open Source licensing issues. The upshot is that embedded developers need to be aware of license issues surrounding both Open Source and proprietary software. Of necessity, embedded software is often intimately tied to the operating system, and includes elements derived or acquired from other sources. While no one expects embedded engineers to be intellectual property attorneys, it is nevertheless essential to understand the license terms of the software you use and create, to be sure that all the elements "play nicely" together.

And the issue cuts both ways. There are also efforts to identify violations of the GPL. The intent here is not the make money, but to defend the integrity of the GPL by putting pressure on violators to clean up their act. In particular, the GPL Violations Project has "outed" a dozen or so embedded Linux vendors who appear to have played fast and loose with the GPL terms. According to Harald Welte, founder of the GPL Violations Project, the most common offenders are networking devices such as routers, followed by set-top boxes and vehicle navigation systems.

Open source licensing expert Bruce Perens has observed that embedded developers seem to have a mindset that "this is embedded, no one can change the source—so the GPL must not really apply to us." It does.

## Alternatives to Linux

Linux isn't the solution to every problem. There are many devices that are simply too resource constrained to run Linux. A majority of the 50,000,000,000 devices expected to

[3] Fear, Uncertainty, and Doubt.

connect to the Internet of Things will likely be too small to run Linux. Nevertheless, these devices can benefit from running an operating system. Whereas Linux is a full-blown operating system with a kernel, device driver model, file systems, network stack, and a large collection of utilities, many alternatives are really just the kernel that handles scheduling, synchronization, and interrupts, and perhaps manages memory.

There are a great many operating systems out there, both open source and proprietary, but two stand out in my mind as being worth discussing.

### *FreeRTOS*

FreeRTOS is an open source project designed to be small and simple. It was originally written by Richard Barry, who subsequently formed Real Time Engineers Ltd. to provide commercial support. It has been ported to 35 microcontrollers. FreeRTOS is distributed under the GPL version 2 with an exception and a restriction. The exception allows you to retain your application code as proprietary, even when it is linked to the FreeRTOS kernel. The restriction is that "FreeRTOS may not be used for any competitive or comparative purpose, including the publication of any form of run time or compile time metric, without the express permission of Real Time Engineers Ltd." The claim is that "this is the norm within the industry and is intended to ensure information accuracy."

FreeRTOS is also available under a commercial license, called OpenRTOS, with additional features:

- A warranty. The open source version has no warranty.
- Professional technical support. The open source version is supported by a volunteer online community.
- IP infringement protection. Real Time Engineers Ltd. will defend commercial license holders against charges of patent or copyright infringement.
- No requirement to open source your changes to the RTOS kernel.

Another commercial product, SafeRTOS, extends FreeRTOS into the realm of safety critical systems, having been certified under a number of international safety standards. SafeRTOS is qualified for use in medical devices. Both OpenRTOS and SafeRTOS were developed by WITTENSTEIN High Integrity Systems in collaboration with Real Time Engineers Ltd.

FreeRTOS itself is just the kernel. On top of that is a whole ecosystem of packages called FreeRTOS + . These packages, some of which are open source while others are proprietary, support features such as:

- FAT file system
- UDP/IP

- TCP/IP
- SSL and TLS
- File and device I/O
- Board support packages

The kernel itself consists of only 3 or 4 C files, and can take as little as 6 Kbytes of storage. It supports both preemptive and cooperative multitasking. Synchronization mechanisms include semaphores, mutexes, and message queues. A "tickless" mode of operation is available for low power applications.

A number of chip vendors supply their own Windows-based development environments for FreeRTOS, mostly derived from Eclipse. The one I happen to be familiar with is LPCExpresso from NXP Semiconductor.

### MicroC/OS

MicroC/OS is a contraction of Micro-Controller Operating System. It was initially developed in 1991[4] by Jean Labrosse, who subsequently wrote a series of articles in *Embedded Systems Programming* magazine about it, as well as a book, *μC/OS The Real-Time Kernel*. MicroC/OS is not an open source project, although it is free to use for noncommercial projects. Labrosse subsequently started a company, Micrium, Inc., to commercialize MicroC/OS.

Like FreeRTOS, MicroC/OS itself is just the kernel, which ranges in size from 6 KB to 24 KB depending on what features are enabled. Additional packages support various forms of networking, file systems, profiling, and tracing. Synchronization mechanisms include semaphores, mutexes, message queues, and event flags.

Two versions are currently supported: MicroC/OS II and MicroC/OS III. They differ primarily in the scheduler implementation. MicroC/OS II uses a clever, very efficient scheduling algorithm that limits the number of tasks to 255, each of which has a unique priority in the range of 0 to 255. Because each task has a unique priority, there is no notion of round robin scheduling.

MicroC/OS III, by contrast, allows an unlimited number of tasks, limited only by the memory of the target system, with an unlimited range of priorities. It also supports round robin scheduling among tasks of equal priority.

A personal note. I first encountered MicroC/OS, at that time called μCOS, in Jean's three-part series in *Embedded Systems Programming*. After downloading and playing around with it, I was so impressed I decided to use it as the basis for a class on real-time programming.

[4] Interestingly, the same year Linus released the first Linux kernel.

I found the source code to be the most complete and readable C code I have ever come across. Every function has an extensive header comment that describes what the function does, what the arguments are, and what the return value is. Symbol names are sufficiently long, and contain enough vowels, to be easily understandable even if the words are not fully spelled out. This is the way code should be written!

Now that we have some idea of the embedded and real-time space, how Linux fits into it, and what the alternatives might be, Chapter 2, Installing linux describes the process of installing Linux on a workstation.

## Resources

Linux resources on the web are extensive. This is a list of some sites that are of particular interest to embedded developers.

linuxfoundation.org – Founded in 2000, the Linux Foundation is a nonprofit consortium dedicated to fostering the growth of Linux. Among its activities are a number of conferences around the world, and a full range of training courses.

embedded.com – The website for what used to be *Embedded Systems Design* magazine. This site is not specifically oriented to Linux, but is quite useful as a more general information tool for embedded system issues.

fsf.org – The Free Software Foundation. While it wasn't called "open source" in the beginning, this is where it all really began.

kernel.org – The Linux kernel archive. This is where you can download the latest kernel versions, as well as virtually any previous version.

opensource.org – The Open Source Initiative (OSI), a non-profit corporation "dedicated to managing and promoting the Open Source Definition for the good of the community." OSI certifies software licenses that meet its definition of Open Source.

osdl.org – Open Source Development Lab, a nonprofit consortium focused on accelerating the growth and adoption of Linux in both the enterprise and, more recently, embedded spaces. In September of 2005, OSDL took over the work of the Embedded Linux Consortium, which had developed a "platform specification" for embedded Linux.

gpl-violations.org – the GPL Violations Project was started to "raise the awareness about past and present violations" of the General Public License. According the website it is "still almost a one-man effort."

sourceforge.net – "World's largest Open Source development website." Provides free services to open source developers including project hosting and management, version control, bug and issue tracking, backups and archives, and communication and collaboration resources.

uclinux.org – The Linux/Microcontroller project is a port of Linux to systems without a Memory Management Unit (MMU).

free-electrons.com – An embedded Linux consulting and training company with a large collection of free training materials.

elinux.org – This is a very extensive wiki with a large number of links to other resources.

slashdot.org – "News for nerds, stuff that matters." A very popular news and forum site focused on open source software in general and Linux in particular.

*Alternatives to Linux*

freertos.org – the site for FreeRTOS

micrium.com – the site for MicroC/OS

CHAPTER 2

# Installing linux

## Chapter Outline



*If Bill Gates had a nickel for every time Windows crashed... Oh wait, he does.*

***Spotted on Slashdot.org***

While it is possible to do embedded Linux development under Windows, it is not easy. Basic Windows lacks many of the tools and features that facilitate embedded development. And why would you want to do that anyway? So we'll keep our focus firmly on Linux, and not delve into the process of setting up a Windows development system.

Even though this book is not an introduction to Linux, it is worth taking a little time to review the installation process and alternative configurations. If you already have a Linux installation that you're happy with, you can probably skip this chapter unless you want to learn about virtualization or dual-booting.

The instructions and steps in this chapter are primarily oriented toward CentOS, but the general concepts apply to pretty much any Linux installation.

Linux for Embedded and Real-time Applications.
DOI: http://dx.doi.org/10.1016/B978-0-12-811277-9.00002-X

## Distributions

Linux installation has improved substantially over the years, to the point that it is a reasonably straightforward, just about painless process. The easiest, and indeed the only sensible, way to install Linux is from a *distribution*, or "distro" for short. There are several hundred distributions floating around on the net, many of which address specific niches such as embedded.

A distribution contains just about everything you would need in your Linux installation, including the kernel, utility programs, graphical desktop environments, development tools, games, media players, and on and on. Most, but not all, distributions make use of a *package manager*, software that combines the individual components into easily managed packages. Fedora and Red Hat distributions use RPM, the Red Hat Package Manager. Debian and its derivatives such as Ubuntu use dpkg. A package contains the software for the component, i.e., the executables, libraries, and so on, plus scripts for installing and uninstalling the component, and dependency information showing what other packages this one requires.

A typical distro may incorporate several thousand packages. In fact, this becomes a marketing game among the various projects and distributors of Linux. "My distro has more packages than your distro." Here are some of the more popular and user friendly distributions, any of which are suitable for embedded development. Web locations are given in the Resources section at the end of the chapter:

### Linux Mint

According to its website, "Linux Mint is the most popular desktop Linux distribution and the 3rd most widely used home operating system behind Microsoft Windows and Apple Mac OS." Well-known open source blogger Steven J Vaughan-Nichols claims in a July 2016 ZDNet article that Linux Mint is the "best of all Linux desktops." "It's not only the best Linux desktop, it's the best desktop operating system—period." techmint.com claims that Mint was the most popular distribution in 2015[1].

Mint is based on Debian and Ubuntu. The Mint team's stated objective is "to produce a modern, elegant and comfortable operating system which is both powerful and easy to use." It seems to be focused on the consumer desktop, with extensive support for multimedia. The latest release is 18 (Sarah), released in August 2016.

[1] http://www.tecmint.com/10-top-most-popular-linux-distributions-of-2015/

### *Debian GNU/Linux*

The Debian Project is an all-volunteer effort comprising around 1000 active developers around the world, supported entirely by donations. The website DistroWatch.com ranks Debian as second in terms of number of page hits per day. In first place is a distro called mint. Debian is one of the oldest distributions, having been first released in 1993.

Debian is known for its "abundance of options." The current stable release, 8.5 (Jesse), has access to online repositories with over 40,000 packages supporting 10 computer architectures, including the usual Intel/AMD 32- and 64-bit processors as well as ARM and IBM eServer zSeries mainframes. Debian claims to be one of the first distributions to use a robust package management system, which is different from RPM. Debian packages get the file extension .deb.

Debian tends to have a longer and perhaps less consistent release cycle than many of the other popular distributions, typically taking 2 years between major releases.

### *Ubuntu*

According to its website, Ubuntu is an ancient African word meaning "humanity to others." It also means "I am what I am because of who we all are." It is a fork of the Debian code base first released in 2004 with the aim of creating an easy to use version of Linux. It is available in two versions, desktop and server.

Each release has a version number consisting of the year and month of release, e.g., 16.04 was released in April of 2016. Releases are made on a predictable 6 month schedule, with each fourth release getting long-term support (LTS). LTS releases have been supported for 3 years for the desktop version, and 5 years for the server version. With the pending release of version 12.04, desktop support will be extended to 5 years.

Historically, Ubuntu, like most other Linux distros, supported both the GNOME and KDE graphical desktop environments. Release 11.04 in spring 2011 introduced a new desktop environment called Unity that is a "shell interface" for GNOME. The intention is to make more efficient use of space on the limited size screens of notebooks and tablets. Some users have criticized the new interface as "too different from and less capable than GNOME," while others find the minimalist approach more appealing than the older paradigm.

While Ubuntu emphasizes ease of use, the developers have made basic system configuration more difficult by eliminating many of the graphical dialogs that support setup and configuration in, e.g., Fedora. Furthermore, the root user account is "locked" so that it is not possible to directly log in as root. You must use the `sudo` command. This apparently is intentional, in the tradition of Windows, as a way of discouraging "average users" from messing around with their systems.

Difficulty in configuration will become an issue in Chapter 4, The host development environment, when we have to change some network parameters. I don't recommend Ubuntu unless it's your favorite distribution, and you're very comfortable with it.

Ubuntu was the third most popular distribution in 2015.

### Red Hat Enterprise Linux

Red Hat Enterprise Linux (RHEL) is specifically targeted at the commercial market, including mainframes. There are server versions for x86, x86-64, Itanium, PowerPC, and IBM System z, and desktop versions for x86 and x86-64. While RHEL derives many of its features from Fedora, it has a more conservative release schedule, with major releases appearing about every 2 years. As of late 2016, the latest stable release is 7.2.

Even though RHEL is a commercial product, it is based entirely on Open Source code. Consequently, Red Hat makes the entire source code base available for download. Several groups have taken advantage of this to rebuild their own versions of RHEL. One of the best known of these is CentOS, said to be the 8th most popular distribution, as of late 2015. These rebuilds remove any reference to Red Hat's trademarks, and point the update systems at non-Red Hat servers. Otherwise, they are functionally identical.

While the rebuilds are free, they are of course not eligible for any kind of Red Hat support.

### Fedora

The Fedora Project was started in late 2003 when Red Hat discontinued its retail line of Linux distributions to focus on enterprise software. The original Fedora release was based on Red Hat Linux 9. Red Hat continues to sponsor the Fedora Project, which serves as a launching point for releases of RHEL.

Fedora prides itself on being on the leading edge of open source software development and, as a result, it has a relatively aggressive release schedule, about every 6 months. As of July 2016, the current release is 24. Having started with Red Hat, Fedora was my favorite Linux distribution up to version 17. I tend to stay one or two major releases behind the leading edge, and only update about every two or three releases. I don't have to have the absolute latest and greatest[2]. Hey, if it works, stick with it.

[2] And neither do you, by the way. The basic functionality doesn't change that much from one release to the next. Much of it is cosmetic. If you're just starting out, pick a fairly recent release and stick with it for a while.

Fedora only runs on Intel/AMD platforms.

I have two basic complaints about new releases. This may also apply to distros other than Fedora:

1. Features tend to "move around." I'm perhaps sensitive to this, because I'm inclined to use the graphical dialogs for configuration and setup. It seems like with each new release, the path to these dialogs changes.
2. Each successive release seems to get more "Windows-like." That is, more menus turn into cutsie icons rather than simple, straightforward text selections. This no doubt reflects a desire to appeal to a more "consumer" audience, but as a computer professional, I find it off-putting. Fortunately, at least for now, there are ways to revert back to the classical menu styles.

I stuck with Fedora 17 long after it was officially obsolete, because I ran into problems with later versions that I never could adequately solve. I subsequently shifted over to CentOS.

### CentOS

CentOS is an abbreviation for Community Enterprise Operating System. It is a free and open source version of RHEL. It is derived from the same open source code as RHEL, with Red Hat's branding and logos changed.

In January 2014, Red Hat announced that it would sponsor the CentOS project, "helping to establish a platform well-suited to the needs of open source developers that integrate technologies in and around the operating system." As a result of these changes, ownership of CentOS trademarks was transferred to Red Hat, which now employs most of the CentOS head developers; however, they work as part of the Red Hat's Open Source and Standards team, which operates separately from the RHEL team.

The latest release is 7.2-1511, December 2015. As of release 7.0, CentOS only officially supports the x86-64 architecture. Unofficial ports are available for alternate architectures. Version 7 will have 10 years of support until 2024.

In July 2010, CentOS overtook Debian to become the most popular Linux distribution for web servers, with almost 30% of all Linux web servers using it, according to w3techs.com. Debian subsequently regained the lead in January 2012.

While just about everything in this book is "distribution-agnostic," one area where distros tend to differ is in setup and configuration. We'll confront that issue in Chapter 4, The host development environment, when we need to make some changes to networking. The descriptions there will emphasize CentOS. With that in mind, if you haven't yet settled on a favorite Linux distribution, I would recommend CentOS, at least while you're working through the book.

### SUSE

SuSE was originally developed in Germany, with the initial release in 1994 making it the oldest existing commercial distribution. The name is a German acronym for Software und System Entwicklung (Software and Systems Development). The name was subsequently changed to SUSE, and is no longer considered an acronym.

Novell acquired SUSE Linux AG in 2003, and in 2005 announced the openSUSE project to allow outside developers to participate. SUSE Linux is available in two forms: openSUSE, driven by the openSUSE project; and SUSE Linux Enterprise, a commercial version. Much like Fedora, openSUSE is on the "bleeding edge" of Linux development with an aggressive release schedule, while SUSE Linux Enterprise sticks to a more conservative schedule.

## Hardware Requirements

Having selected a distribution, you need something to run it on. Any modern PC will work just fine as a development host. Minimum requirements are: a Pentium class processor, 1 GB of RAM for graphical operation, and at least 20 Gbytes of disk for a "workstation" class Linux installation. Of course, more RAM and disk is always better.

While you can "get by" with 20 GB of disk, later we'll be building the Yocto project, which requires more like 50 GB. I do most of my Linux work on a "virtual" machine (more about that later) running under Windows 7 on an HP laptop. The virtual machine gets 1 GB of RAM and has a 65 GB disk. I currently run CentOS 7.

You will need at least one asynchronous serial port. A USB-to-serial converter should work fine. You will also need a network interface. We'll use a combination of serial and network ports to communicate with the target, as well as to debug target code.

## Installation Scenarios

The next decision then is how you want to install Linux. There are basically three installation scenarios.

### Stand-Alone

This is the obvious choice if you can dedicate a machine to Linux. You will let the installation process format the entire disk.

### Dual-Booting

In many cases, though, you'll probably want to install Linux on a machine that already runs some variant of Windows. There are a couple of approaches to that. This section describes dual-booting, the next one describes virtualization.

In the dual-boot scenario, you select at boot time which operating system to boot. That OS takes full control of the machine. The Linux installation will replace the standard Windows boot loader with GRUB, the GRand Unified Bootloader. GRUB then offers the option of selecting the OS to boot, as shown in Fig. 2.1.

While historically dual-booting has been the most popular installation scenario for average users, it is also the most complicated, because it requires reconfiguring your hard disk. The most common case is you already have a version of Windows installed on the machine, and you want to add Linux.

At this point Windows probably occupies the entire disk, and so you have to make space available for Linux. Fundamentally, this requires reducing the size of the disk partition holding Windows so you can create unallocated space for Linux. Remember that you'll need on the order of 20 GB of disk space for Linux, so your disk must have at least that

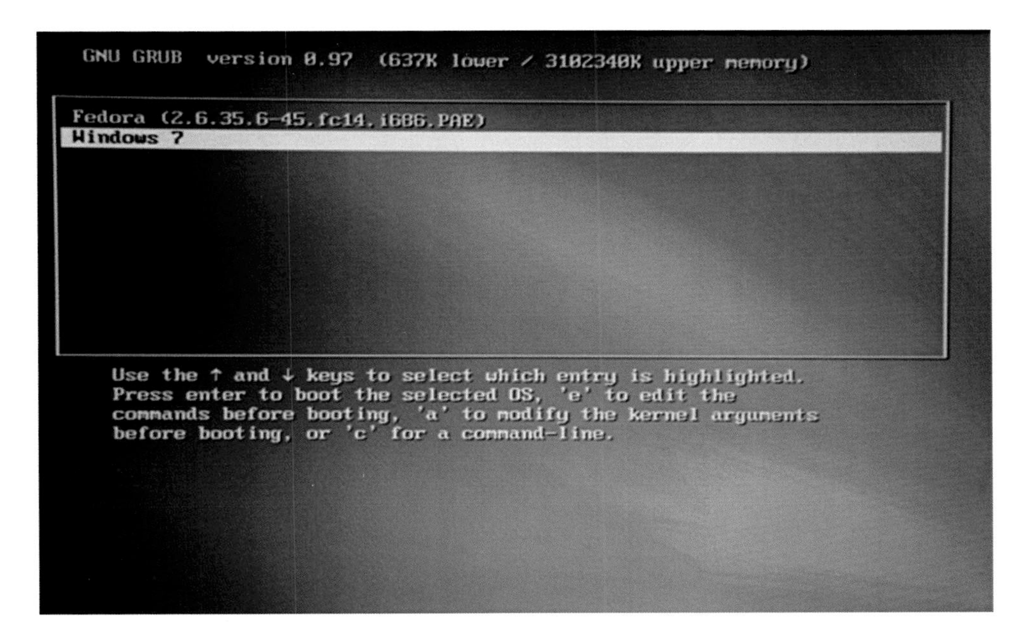


**Figure 2.1**
GRUB boot menu.

much contiguous free space. You can use the Windows defragmentation tool to put all free space at the "end" of the disk.

Windows 7 has a slick disk management tool that handles partitioning. From the Start menu, select Control Panel > Administrative Tools > Computer Management. In the Computer Management dialog, select Storage > Disk Management. That displays the menu shown in Fig. 2.2. This shows five partitions, the largest of which is identified as drive C:.

A little background on disk partitioning is in order here. In the DOS/Windows/PC world, a disk can have up to four *primary* partitions. Any of these primary partitions can be designated an *extended* partition, which in turn can hold several *logical* partitions. There's no fixed upper limit on the number of logical partitions an extended partition can hold, but owing to the way in which Linux accesses partitions, 12 is the practical upper limit on a single disk drive.

Take a look at the partition list in Fig. 2.2, and compare it with the list in Listing 2.1 derived from the Linux `fdisk` command on the same disk. This particular machine is already configured for dual booting. Note that `fdisk` reports six partitions, whereas the Windows Disk Manager only shows five. `/dev/sda4` is identified as an extended partition.

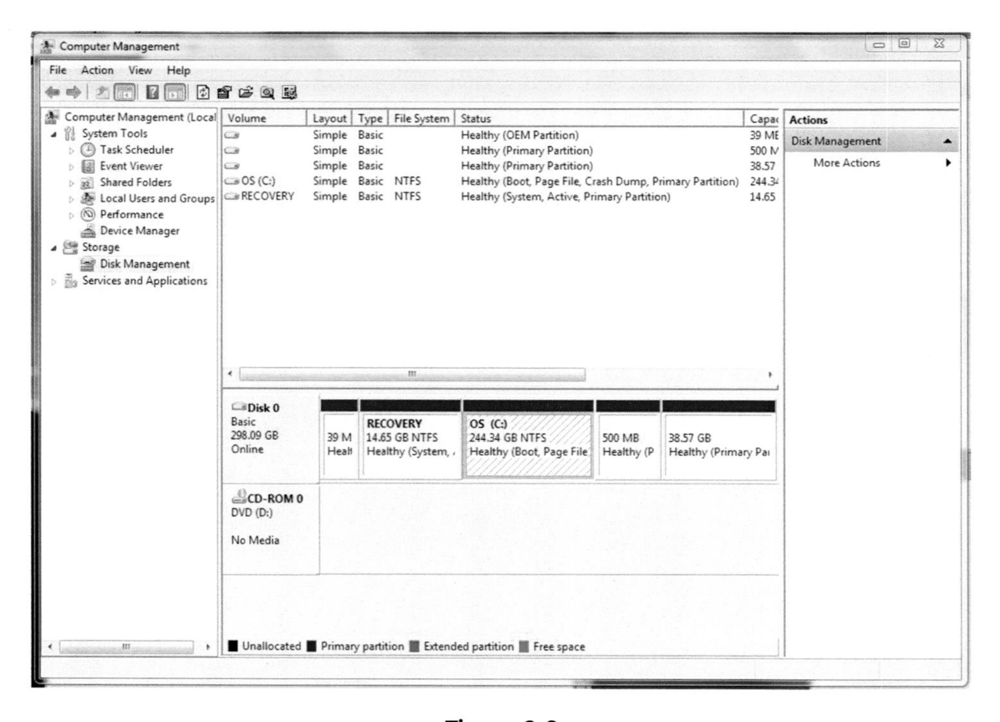


**Figure 2.2**
Windows 7 Disk Manager.

```
   Device Boot      Start        End      Blocks   Id  System
/dev/sda1              63      80324       40131   de  Dell Utility
/dev/sda2   *       81920   30801919    15360000    7  HPFS/NTFS
/dev/sda3        30801920  543220399   256209240    7  HPFS/NTFS
/dev/sda4       543221760  625141759    40960000    5  Extended
/dev/sda5       543223808  544247807      512000   83  Linux
/dev/sda6       544249856  625141759    40445952   8e  Linux LVM
```

**Listing 2.1**
`fdisk` output.

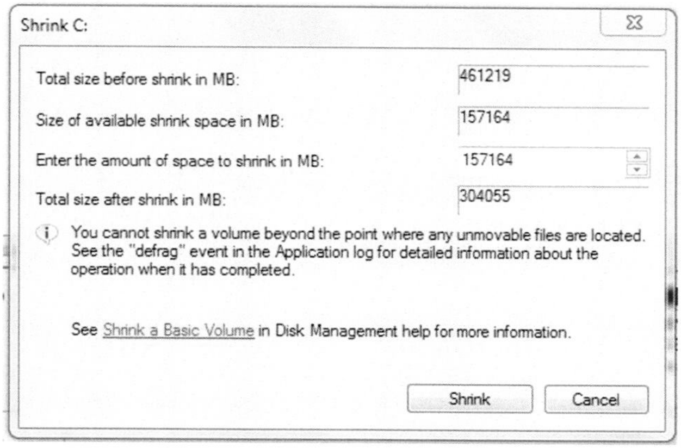

**Figure 2.3**
Shrink partition.

`sda5/` and `sda6/` are logical partitions within `sda4/`. The Disk Manager doesn't show the extended partition.

Oddly, the Disk Manager identifies all five partitions as primary, but `fdisk` shows us that the two on the right of the graphic representation are in fact logical partitions within an extended partition.

With that background, it's time to create some empty space for Linux. But first, before you make any changes to a disk drive, **Always back up your data.** Even though these tools are supposed to work, stuff happens. So be prepared. Right-click on the OS (C:) partition, and select `Shrink volume`... Note incidentally that Windows tends to call partitions volumes. The Disk Manager queries the partition to see how small it can be made. This can take a while on a large disk. Finally, the dialog of Fig. 2.3 appears. While the labels are a little confusing, this is telling us that 113 MB can be removed from the C: partition. Clearly, you want to leave some space free on the C: drive, so you wouldn't want to take all 113 MB.

Subsequently, the Linux installation will format only the partition(s) that will be used by Linux.

### Virtualization

But the big buzzword in computing these days is virtualization, the process of running one operating system on top of another. The base, or native, operating system is called the *host*. It runs a *virtual machine manager*, VMM, that in turn runs one or more virtual machines called *guests*. Whereas in the dual-boot scenario one or the other operating system is running exclusively, with virtualization the host and the guests are running simultaneously. You can even move files and data between them seamlessly.

There are two popular VMMs—Vmware and VirtualBox. VMware is a commercial product from a company of the same name. They offer a free version called VMware Player that runs on both Windows and Linux hosts.

VirtualBox is an open source package sponsored by Oracle (formerly Sun Microsystems). It too is available for both Windows and Linux hosts, but also supports Mac OS X and Solaris.

The two packages are fairly similar in their installation and operation. After installing the software (see Resources section for download pages), you create one or more guest machines allocating resources to them, such as disk and RAM. A "wizard" steps you through the process of creating a new machine. A disk in the guest machine is represented by a very large file in the host. Then you install an operating system on the guest, in much the same way you would install it natively.

I happen to use VirtualBox.

### Installing VirtualBox

Installing VirtualBox is much like installing any Windows software. You download and run the installation executable. The default configuration is fine. You'll be prompted if you want to install the Extension Pack. Yes, you do.

The next step then is to create a virtual "guest" machine and install Linux on it. Start the VirtualBox Manager and click the New icon in the upper left. Give the machine a name. VirtualBox will often infer the type and version from the name. Be sure they're correct. Allocate RAM. I have 8 GB in my laptop, and allocate 2 GB to guest machines. I found that allocating more than 2 GB to the guest caused Windows to slow down.

Create a virtual hard disk. Leave the default disk file type, VirtualBox disk image (.vdi). I use dynamically allocated storage for the disk file. Your choice. By default, the disk file name is the same as the machine name. Set the size. I suggest 65 GB to have enough to comfortably build the Yocto project. When you click the Create button, the machine is created.

There are some settings you'll want to change in the guest machine itself.

- In `General>Advanced`, set `Shared Clipboard` and `Drag 'n Drop` to `Bidirectional`.
- In `Network`, change `Attached to:` to `Bridged Adapter`. Make sure the correct adapter is selected. If your workstation has both wired and wireless connections, you might want to enable the second adapter. Then you can connect the target board directly to the workstation through the Ethernet port, and use the wireless port for Internet access. Note that VirtualBox makes all network adapters appear to the guest as an Intel PRO/1000.
- If you are using a USB-to-Serial converter, plug it in. Then in `USB`, click the `Add a USB filter` icon on the right, and select your converter. This allows it to be automatically connected to the guest machine whenever it is plugged in.
- In `Shared Folders` you'll probably want to set up a shared folder so that you can easily move files between the host and the guest. Click the `Add a new shared folder` icon on the right. Browse to the folder you want to share and give it a name. The name becomes the device you mount in the Linux guest.
- When you're ready to boot the new guest and install Linux, go to `Storage` and click the optical disk icon under `Controller: IDE`. Then click the optical disk icon on the far right and select the ISO file for your distro, or the physical drive if you're using a real DVD.

After you get Linux installed and booted, you'll find there are some limitations to the display and mouse in basic VirtualBox. The display size is limited, and the mouse works either in the host or the guest, but not in both. These limitations are overcome by installing the VirtualBox Guest Additions.

If you're new to Linux, the instructions in this section may not make much sense until you've read Chapter 3, Introducing linux. You'll need the GCC compiler package installed (you'll need that anyway as we go along) and the kernel headers package for the kernel version you're running. The kernel headers package should be installed as part of the initial installation.

Click the VirtualBox `Devices` menu and select `Insert Guest Additions CD Image`. As root user, mount the CD image. In a shell window, cd to the just-mounted image and execute:

```
sh ./VBoxLinuxAdditions.run
```

This script builds several kernel modules. Restart the guest machine when the script finishes. You can now set the guest display to full screen if desired, and the mouse works seamlessly between the host and guest.

## DVD or Live CD?

Most Linux distributions are available in at least three forms: a DVD, a collection of CDs (up to about six these days), or a *Live CD.* A Live CD is a minimal bootable Linux system that gives you an opportunity to play around with a new distribution without actually installing it. Of course you can't save any data or modify the configuration. Every time it boots it reverts to the default configuration.

One of the options the Live CD offers is to install the distribution. When you select this option, it runs through essentially the same installation process as you would with a DVD or complete set of CDs. The difference is that packages to install are pulled from the network rather than from physical media.

## Installation Process

Regardless of which scenario or installation medium you choose, the installation process is pretty much the same. This section describes CentOS installation, but other distributions have a very similar process:

- Download the distribution medium (see resources section). This will be a .iso file, an exact image of either a DVD or a Live CD.
- Burn the .iso to the corresponding physical medium. You'll need a disk burning program to transfer the .iso file(s) to physical media. In a virtual machine environment, you can mount the .iso file directly as if it were a CD or DVD.
- Boot the computer from the installation medium or click the install icon from a Live CD. You may have to configure your PCs BIOS to boot from optical media.
- Follow the instructions.

The installation process itself is fairly straightforward. With CentOS, you start by selecting a language, which defaults to United States English. Clicking Continue brings up the `INSTALLATION SUMMARY` dialog shown in Fig. 2.4.

Under `SOFTWARE SELECTION`, I suggest selecting `KDE Plasma Workspaces` as your base environment. That's only because I happen to prefer the KDE graphical desktop environment over GNOME and, to the extent that the remainder of the book displays graphical desktop images, they will be KDE. I just happen to like the KDE presentation better. It's more "Windows-like." The truth is, I do most of my day-to-day computing work in Windows (hey, I'm writing this book in Word). If you've spent much of your career in front of a Windows screen, I think you'll find KDE more to your liking. But if you have a strong preference for GNOME, by all means, select it or the `Development and Creative Workstation`. At a minimum, you should select `Development Tools` under Add-Ons. If you're going with the KDE Plasma Workspace you might also want to select `KDE Applications`.

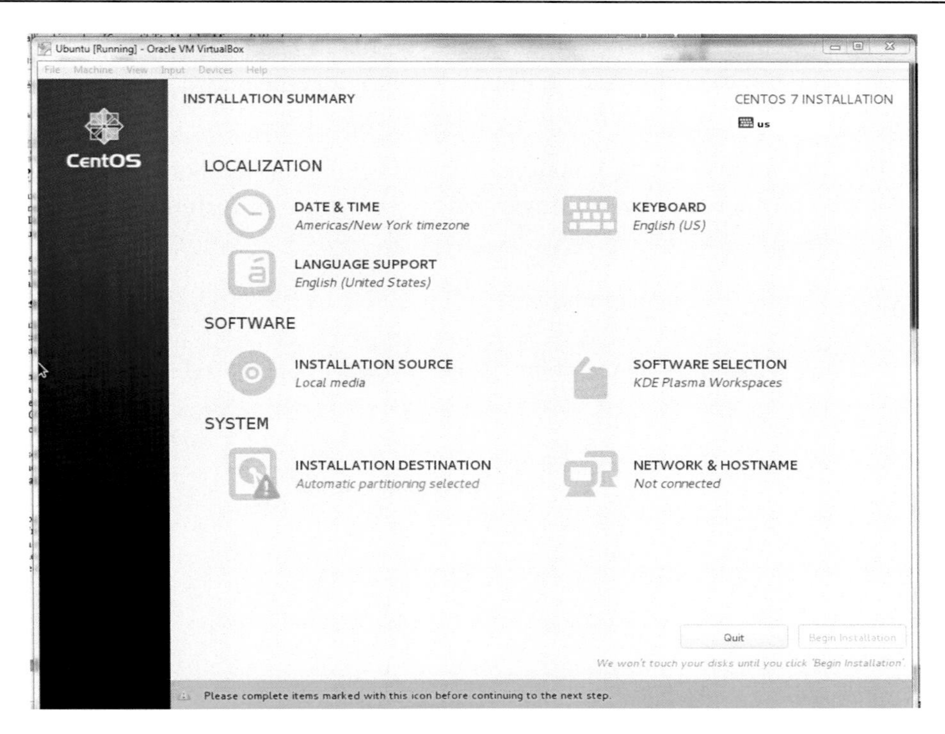


**Figure 2.4**
Main setup menu.

Under NETWORK & HOST NAME, select what may be the only network interface. Mine is called Ethernet (enp0s3). Turn it on and then click Configure... In the General tab, it's probably a good idea to select Automatically connect to this network when it is available.

Finally, select INSTALLATION DESTINATION. Select what is probably the only disk drive visible, then click Done.

Click Begin Installation and you are asked to set the root password and create a normal user. The root user is the system administrator who can do anything. If you create a password that CentOS considers too weak, you'll have to click Done twice. I know it's bad practice, but I use the same password for root and my normal user. I don't have anything sensitive on my Linux machine.

Go through the Create User dialog. You can even create a normal user without a password if you choose. Up to you. While you've been doing this, the installation has been proceeding apace. When the installation completes, remove the optical media and click Reboot. When CentOS boots up, you'll be asked to accept the license. It's pretty simple as EULAs go.

There's one more step before CentOS finally lets you log in and start up the graphical desktop. You are asked to enable and configure Kdump, a kernel crash dumping mechanism. Your call. I've never used it but I leave it enabled.

Following that, you'll have a fully functional Linux installation. Reboot it when you're ready for the next chapter where we'll review many of the useful features of Linux.

## Resources

iso.linuxquestions.org – This site has downloadable ISO image files for close to 500 Linux distributions.

*Specific Distribution Sites*
linuxmint.com
debian.org
ubuntu.com
redhat.com
fedoraproject.org
centos.org
opensuse.org

*Other Resources*
vmware.com – Information about the VMware virtual machine manager. Download VMware Player here.
virtualbox.org – Site for the VirtualBox VMM.

# Introducing linux

## Chapter Outline



> *There are two major products to come out of Berkeley: LSD and Unix. We don't believe this to be a coincidence*
>
> ***Jeremy S. Anderson***

For those who may be new to Unix-style operating systems, this chapter provides an introduction to some of the salient features of Linux, especially those of interest to embedded developers. This is by no means a thorough introduction, and there are many books available that delve into these topics in much greater detail.

Linux for Embedded and Real-time Applications.
DOI: http://dx.doi.org/10.1016/B978-0-12-811277-9.00003-1

Feel free to skim, or skip this chapter entirely, if you are already comfortable with Unix and Linux concepts.

## Running Linux—KDE

Boot up your Linux machine and log in as your normal user. If you are running the K Desktop Environment (KDE) on CentOS 7, you'll see the screen shown in Fig. 3.1. Unless you are running VirtualBox or VMware, you won't see the top two lines and the very bottom menu line. If you are running a different version of Linux, your screen will probably look different, but should have most of the same features.

At the bottom left is a menu icon that is the CentOS logo. It serves the same purpose as the Start Menu in Windows. KDE calls this the "Application Launcher Menu." Initially, clicking this icon brings up a set of cutsie icon-based menus that I personally find difficult to use. Fortunately you can change it to the older list-based style. Right-click the Application Launcher and select `Switch to Classic Menu Style`. Recognize, of course, that this is purely a matter of personal taste.

**Figure 3.1**
Initial KDE screen.

### File Manager

One of the first things I want to do with a new system is open a file manager so I can see what is on the system. Click the Application Launcher and select `Home`. The file manager initially displays the contents of your *home directory*, the place where you store all of your own files. Note that, by default, KDE uses a single click to activate items. You can change that to double click by selecting `Settings > System Settings` from the Application Launcher. Scroll down to `Hardware` and select `Input Devices > Mouse`.

The default file manager in recent Red Hat-based releases is called Dolphin and, again, is not my cup of tea. I find the older file manager, Konqueror, to be easier to use and to provide a more useful presentation. You can change file managers by right-clicking on the Application Launcher and selecting `Edit Applications...` In the KDE Menu Editor, select `Home` and enter the Command: shown in Fig. 3.2.

Fig. 3.3 shows Konqueror as I have it configured. Not surprisingly, that is not how it looks initially. To get the navigation panel on the left, click on the red "Root folder" icon on the far left. There are lots of options in Konqueror, so play around with it to get it exactly to your liking.

You probably won't see much in your home directory initially. That's partly because many of the entries are "hidden," that is not visible to the file manager. By convention, any file or directory whose name begins with a dot is considered to be hidden. There's an option in the `View` menu to `Show hidden files`.

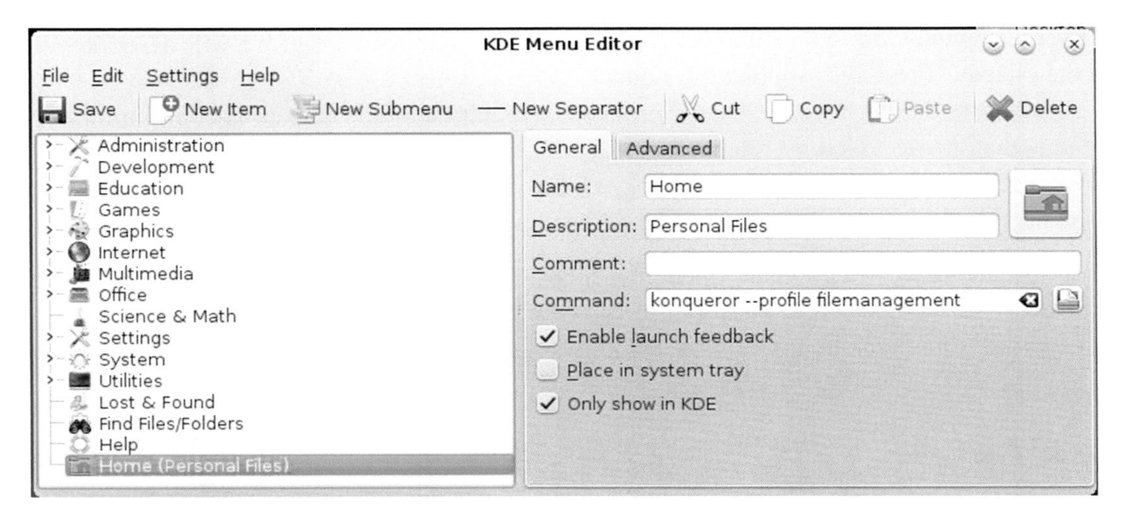


**Figure 3.2**
KDE Menu Editor.

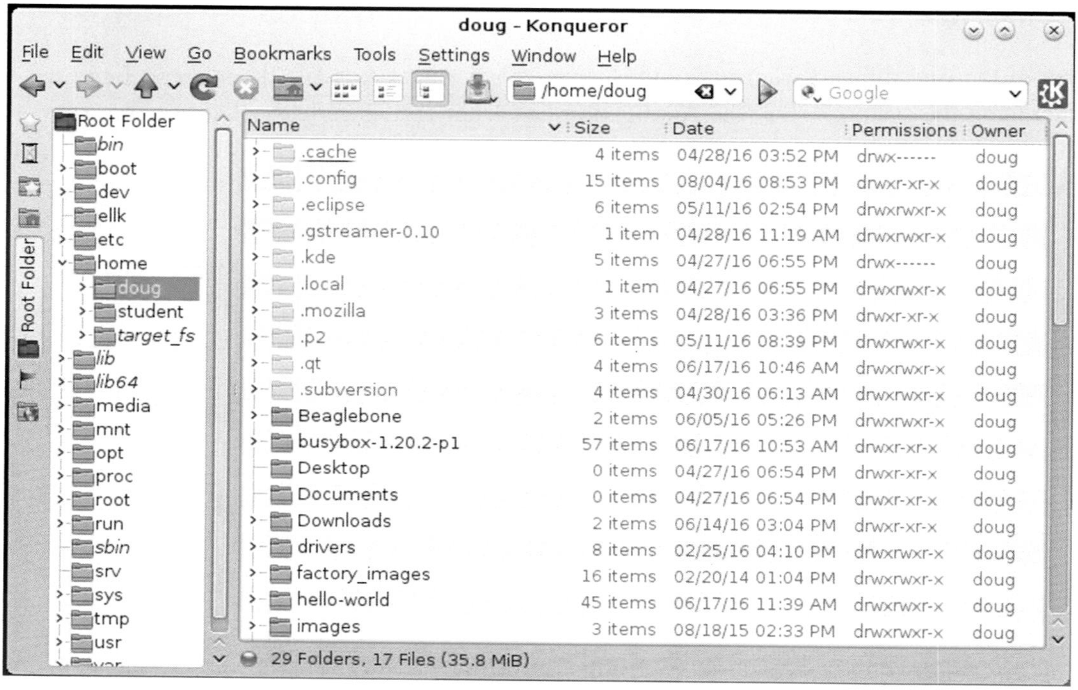

Figure 3.3
Konqueror file manager.

### Shell Window

The other window that you'll be using a lot is the command shell that we will describe later in the chapter. From the Application Launcher Menu, select `System>Konsole (Terminal)`. Fig. 3.4 shows how I have the shell configured. Again, there are numerous configuration options that are accessed by selecting `Configure Profiles...` from the `Settings` menu. I happen to like black text on a white background, and I set the size to 80 × 25, because that matches the old serial CRT terminals that I used when I was starting out in this business.

The `File` menu offers options to open new shell windows and multiple tabs within a shell window. It is not unusual to have two or three shell windows open simultaneously, and to have two or three tabs in any of those windows.

## Linux Features

Here are some of the important features of Linux, and Unix-style operating systems in general.

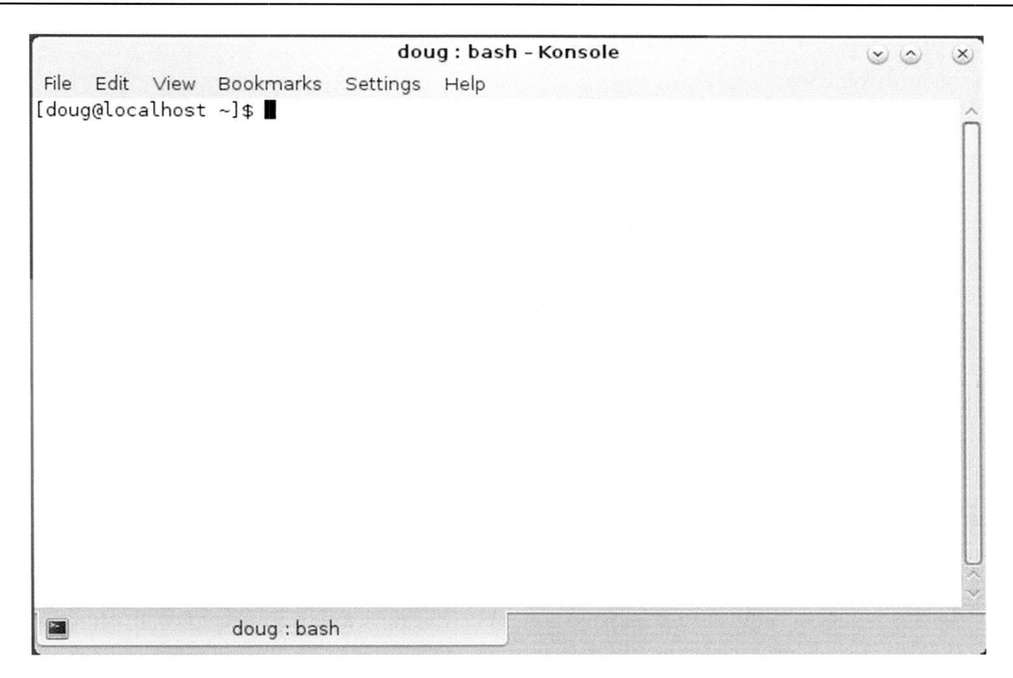


**Figure 3.4**
Command shell window.

- Multitasking. The Linux scheduler implements true, preemptive multitasking, in the sense that a higher priority process made ready by the occurrence of an asynchronous event will preempt the currently running process. But while it is preemptible[1], there are relatively large latencies in the kernel that make it generally unsuitable for hard real-time applications, although that has improved substantially over the years and we'll go into that later on. Additionally, the default scheduler implements a "fairness" policy that gives all processes a chance to run.
- Multiuser. Unix evolved as a time-sharing system that allowed multiple users to share an expensive (at that time anyway) computer. Thus, there are a number of features that support privacy and data protection. Linux preserves this heritage and puts it to good use in server environments.[2]

[1] Is it "preemptible" or "preemptable"? Word 2010's spelling checker says they're both wrong. The only online reference to this debate I could find is a thread on Yahoo Answers from 2008 that comes down on the side of preemptible. I think I'll stick with that.

[2] Although my experience in the embedded space is that the protection features, particularly file permissions, can be a downright nuisance. Some programs, the Firefox browser is an example, don't have the courtesy to tell you that you don't have permission to write the file, they just sit there and do nothing.

- Multiprocessing. Linux offers extensive support for true symmetric multiprocessing where multiple processors are tightly coupled through a shared memory bus. This has become particularly significant in the era of multicore processors.
- Protected memory. Each Linux process operates in its own private memory space, and is not allowed to directly access the memory space of another process. This prevents a wild pointer in one process from damaging the memory space of another process. The errant access is trapped by the processor's memory protection hardware, and the process is terminated with appropriate notification.
- Hierarchical file system. Yes, all modern operating systems—even DOS—have hierarchical file systems. But the Linux/Unix model adds a couple of nice wrinkles on top of what we are used to with traditional PC operating systems:
    - Links. A link is simply a file system entry that points to another file, rather than being a file itself. Links can be a useful abstraction mechanism, and a way to share files among multiple users. They also find extensive use in configuration scenarios for selecting one of several optional files.
    - Device-independent I/O. Again, this is nothing new, but Linux takes the concept to its logical conclusion by treating every peripheral device as an entry in the file system. From an application's viewpoint, there is absolutely no difference between writing to a file and writing to, say, a printer.

## Protected Mode Architecture

Before getting into the details of Linux, let's take a short detour into protected mode architecture. The implementation of protected mode memory in contemporary Intel processors first made its appearance in the 80386. It utilizes a full 32-bit address for an addressable range of 4 GB. Access is controlled such that a block of memory may be: Executable, Read only, or Read/Write. Current Intel 64-bit processors implement a 48-bit address space.

The processor can operate in one of four *Privilege Levels*. A program running at the highest privilege level, level 0, can do anything it wants—execute I/O instructions, enable and disable interrupts, modify descriptor tables. Lower privilege levels prevent programs from performing operations that might be "dangerous." A word processing application probably should not be messing with interrupt flags, for example. That's the job of the operating system.

So application code typically runs at the lowest level, while the operating system runs at the highest level. Device drivers and other services may run at the intermediate levels. In practice, however, Linux and most other operating systems for Intel processors only use levels 0 and 3. In Linux, level 0 is called "Kernel Space" while level 3 is called "User Space."

## Real Mode

To begin our discussion of protected mode programming in the x86, it's useful to review how "real" address mode works.

Back in the late 1970s, when Intel was designing the 8086, the designers faced the dilemma of how to access a megabyte of address space with only 16 bits. At the time a megabyte was considered an immense amount of memory. The solution they came up with, for better or worse, builds a 20-bit (1 megabyte) address out of two 16-bit quantities called the Segment and Offset. Shifting the segment value four bits to the left and adding it to the offset creates the 20-bit linear address (see Fig. 3.5).

The x86 processors have four segment registers in real mode. Every reference to memory derives its segment value from one of these registers. By default, instruction execution is relative to the Code Segment (CS), most data references (MOV for example) are relative to the data segment (DS), and instructions that reference the stack are relative to the Stack Segment (SS). The Extra Segment is used in string move instructions, and can be used whenever an extra DS is needed. The default segment selection can be overridden with segment prefix instructions.

A segment can be up to 64 Kbytes long, and is aligned on 16-byte boundaries. Programs less than 64 Kbytes are inherently position-independent, and can be easily relocated anywhere in the 1 Mbyte address space. Programs larger than 64 Kbytes, either in code or data, require multiple segments and must explicitly manipulate the segment registers.

## Protected Mode

Protected mode still makes use of the segment registers, but instead of providing a piece of the address directly, the value in the segment register (now called the *selector*) becomes an index into a table of *Segment Descriptors*. The segment descriptor fully describes a block of memory including, among other things, its base and limit (see Fig. 3.6). The linear address

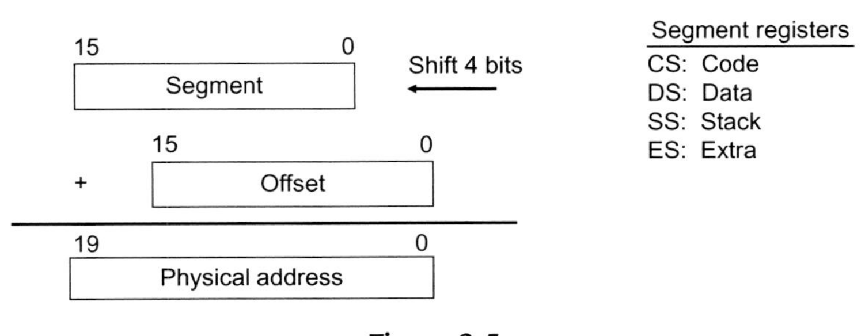


**Figure 3.5**
Real mode addressing.

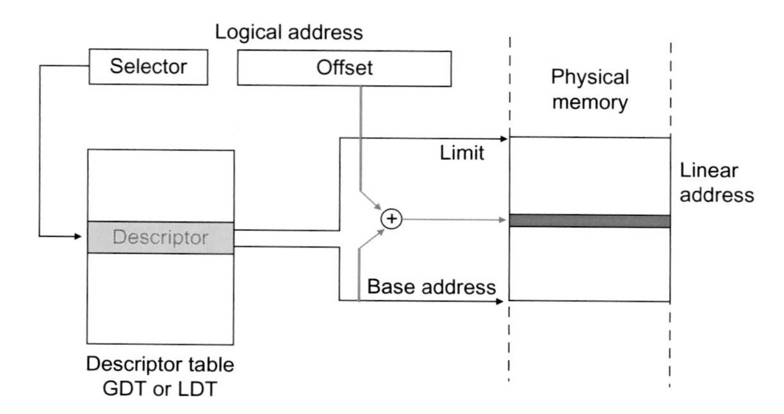


**Figure 3.6**
Protected mode addressing.

in physical memory is computed by adding the offset in the logical address to the base contained in the descriptor. If the resulting address is greater than the limit specified in the descriptor, the processor signals a memory protection fault.

A descriptor is an 8-byte object that tells us everything we need to know about a block of memory.

Base Address[31:0]: Starting address for this block/segment.
Limit[19:0]: Length of this segment. This may be either the length in bytes (up to 1 Mbyte), or the length in 4 Kbyte *pages*. The interpretation is defined by the Granularity bit.
Type: A 4-bit field that defines the kind of memory that this segment describes
S 0 = this descriptor describes a "System" segment. 1 = this descriptor describes a code ordata segment.
DPL Descriptor Privilege Level: A 2-bit field that defines the minimum privilege level required to access this segment.
P Present: 1 = the block of memory represented by this descriptor is present in memory. Used in paging.
G Granularity: 0 = Interpret Limit as bytes. 1 = Interpret Limit as 4 Kbyte pages.

Note that, with the Granularity bit set to 1, a single segment descriptor can represent the entire 4 Gbyte address space.

Normal descriptors (S bit = 1) describe memory blocks representing data or code. The Type field is four bits, where the most significant bit distinguishes between Code and Data segments. Code segments are executable, data segments are not. A CS may or may not also be readable. A data segment may be writable. Any attempted access that falls outside the scope of the Type field—attempting to execute a data segment for example—causes a memory protection fault.

### "Flat" Versus Segmented Memory Models

Because a single descriptor can reference the full 4 Gbyte address space, it is possible to build your system by reference to a single descriptor. This is known as "flat" model addressing and is, in effect, a 32-bit equivalent of the addressing model found in most 8-bit microcontrollers, as well as the "tiny" memory model of DOS. All memory is equally accessible, and there is no protection.

Linux actually does something similar. It uses separate descriptors for the operating system and each process so that protection is enforced, but it sets the base address of every descriptor to zero. Thus, the offset is the same as the virtual address. In effect, this does away with segmentation.

### Paging

Paging is the mechanism that allows each task to pretend that it owns a very large flat address space. That space is then broken down into 4 Kbyte *pages*. Only the pages currently being accessed are kept in main memory. The others reside on disk.

As shown in Fig. 3.7, paging adds another level of indirection. The 32-bit linear address derived from the selector and offset is divided into three fields. The high-order 10 bits serve as an index into the *Page Directory*. The Page Directory Entry points to a *Page Table*. The next 10 bits in the linear address provide an index into that table. The Page Table Entry (PTE) provides the base address of a 4 Kbyte page in physical memory called a *Page*

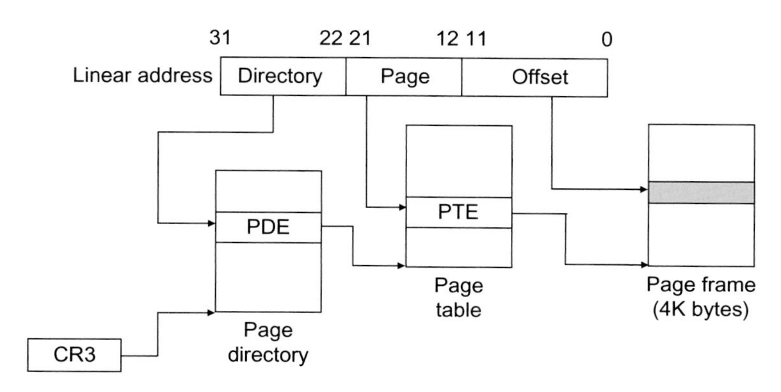


**Figure 3.7**
Paging.

*Frame*. The low order 12 bits of the original linear address supplies the offset into the page frame. Each task has its own Page Directory pointed to by processor control register CR3.

At either stage of this lookup process, it may turn out that either the Page Table or the Page Frame is not present in physical memory. This causes a *Page Fault*, which in turn causes the operating system to find the corresponding page on disk and load it into an available page in memory. This in turn may require "swapping out" the page that currently occupies that memory.

A further advantage to paging is that it allows multiple tasks or processes to easily share code and data by simply mapping the appropriate sections of their individual address spaces into the same physical pages.

Paging is optional, you don't have to use it, although Linux does. Paging is controlled by a bit in processor register CR0.

Page Directory and Page Table entries are each 4 bytes long, so the Page Directory and Page Tables are a maximum of 4 Kbytes, which also happens to be the Page Frame size. The high-order 20 bits point to the base of a Page Table or Page Frame. Bits 9 to 11 are available to the operating system for its own use. Among other things, these could be used to indicate that a page is to be "locked" in memory, i.e., not swappable.

Of the remaining control bits the most interesting are:

P Present: 1 = this page is in memory. If this bit is 0, referencing this Page Directory or PTE causes a page fault. Note that when P = = 0 the remainder of the entry is not relevant.

A Accessed: 1 = this page has been read or written. Set by the processor but cleared by the OS. By periodically clearing the Accessed bits, the OS can determine which pages have not been referenced in a long time, and are therefore subject to being swapped out.

D Dirty: 1 = this page has been written. Set by the processor but cleared by the OS. If a page has not been written to, there is no need to write it back to disk when it has to be swapped out.

### *64-Bit Paging*

The paging model thus far described is for 32-bit x86 processors. It is described as a 10-10-12 model, because the 32-bit linear address is divided into three fields of, respectively, 10 bits, 10 bits, and 12 bits. In a 64-bit machine, entries in the Page Directory and Page Table are 8 bytes, so a 4 KB page holds 512 entries, or 9 bits. Current 64-bit processors only implement 48 bits of physical addressing for a maximum of 256 TB of memory. Two more tiers of address translation are added to yield a 9-9-9-9-12 model.

## The Linux Process Model

OK, back to Linux. The basic structural element in Linux is a *process* consisting of executable code and a collection of *resources* like data, file descriptors, and so on. These resources are fully protected, such that one process cannot directly access the resources of another. In order for two processes to communicate with each other, they must use the interprocess communication mechanisms defined by Linux, such as shared memory regions or pipes.

This is all well and good as it establishes a high degree of protection in the system. An errant process will most likely be detected by the system and thrown out before it can do any damage to other processes (see Fig. 3.8). But there is a price to be paid in terms of excessive overhead in creating processes and using the interprocess communication mechanisms.

A *thread* on the other hand is code only. Threads only exist within the context of a process, and all threads in one process share its resources. Thus, all threads have equal access to data memory and file descriptors. This model is sometimes called *lightweight multitasking*, to distinguish it from the Unix/Linux process model.

The advantage of lightweight tasking is that interthread communication is more efficient. The drawback, of course, is that any thread can clobber any other thread's data. Historically, most real-time operating systems have been structured around the lightweight model. In recent years, of course, the cost of memory protection hardware has dropped

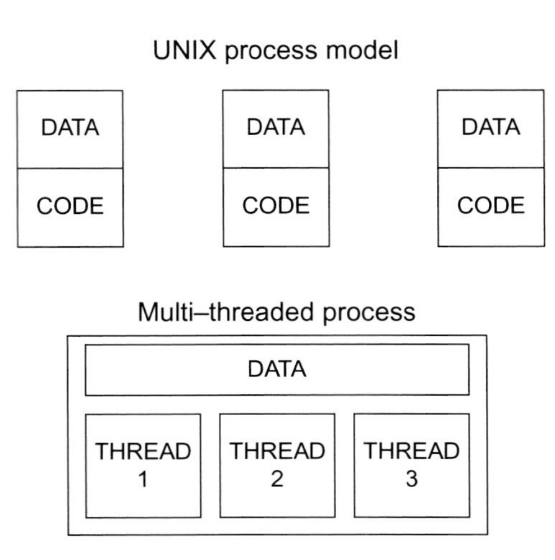


**Figure 3.8**
Processes versus threads.

dramatically. In response, many RTOS vendors now offer protected mode versions of their systems that look like the Linux process model.

### The Fork() Function

Linux starts life with one process, the `init` process, created at boot time. Every other process in the system is created by invoking `fork()`. The process calling `fork()` is termed the *parent*, and the newly-created process is termed the *child*. So every process has ancestors, and may have descendants depending on who created who.

If you have grown up with multitasking operating systems where tasks are created from functions by calling a task creation service, the fork process can seem downright bizarre. `fork()` creates a *copy of the parent process*—code, data, file descriptors, and any other resources the parent may currently hold. This could add up to megabytes of memory space to be copied. To avoid copying a lot of stuff that may be overwritten anyway, Linux employs a *copy-on-write* strategy.

`fork()` begins by making a copy of the process data structure, and giving it a new process identifier (PID) for the child process. Then it makes a new copy of the Page Directory and Page Tables. Initially, the page table entries all point to the same physical pages as the parent process. All pages for both processes are set to read-only. When one of the processes tries to write, that causes a page fault, which in turn causes Linux to allocate a new page for that process and copy over the contents of the existing page.

Since both processes are executing the same code, they both continue from the return from `fork()` (this is what is so bizarre!). In order to distinguish parent from child, `fork()` returns a function value of 0 to the child process, but returns the PID of the child to the parent process. Listing 3.1 is a trivial example of the fork call.

### The Execve() Function

Of course, what really happens 99% of the time is that the child process invokes a new program by calling `execve()` to load an executable image file from disk. Listing 3.2 shows in skeletal form a simple command line interpreter. It reads a line of text from `stdin`, parses it, and calls `fork()` to create a new process. The child then calls `execve()` to load a file and execute the command just entered. `execve()` overwrites the calling process's code, data, and SSs.

If this is a normal "foreground" command, the command interpreter must wait until the command completes. This is accomplished with `waitpid()` which blocks the calling process until the process matching the pid argument has completed. Note, by the way, that most multitasking operating systems do not have the ability to block one process or task pending the completion of another.

```
#include <unistd.h>
#include <>

pid_t pid;

void do_child_thing (void)
{
    printf ("I am the child.  My PID is %d\n", pid);
}
void do_parent_thing (void)
{
    printf ("I am the parent.  My child's PID is %d\n", pid);
}

void main (void)
{
    switch (pid = fork())
    {
        case -1:
            printf ("fork failed\n");
            break;
        case 0:
            do_child_thing();
            break;
        default:
            do_parent_thing();
    }
    exit (0);
}
```

**Listing 3.1**
Trivial example of fork.

If `execve()` succeeds, it does not return. Instead, control is transferred to the newly-loaded program.

## The Linux File System

The Linux file system is in many ways similar to the file system you might find on a Windows PC or a Macintosh. It is a hierarchical system that lets you create any number of subdirectories under a *root directory* identified by "/." Like Windows, file names can be very long. However, in Linux, as in most Unix-like systems, filename "extensions," the part of the filename following ".," have much less meaning than they do in Windows. For example, while Windows executables always have the extension ".exe," Linux executables rarely have an extension at all. By and large, the contents of a file are identified by a file header, rather than a specific extension identifier. Nevertheless, many applications, the C compiler for example, do support default file extensions.

```
#include <unistd.h>

void main (void)
{
        char *argv[10], *filename;
        char text[80];
        char foreground;
        pid_t pid;
        int status;

        while (1)
        {
                gets (text);
// Parse the command line to derive filename and
// arguments. Decide if it's a foreground command.
                switch (pid = fork())
                {
                        case -1:
                                printf ("fork failed \n");
                                break;
                        case 0: // child process
                                if (execve (filename, argv, NULL) < 0)
                                        printf ("command failed \n");
                                break;
                        default: // parent process
                                if (foreground)
                                        waitpid (&status, pid);
                }
        }
}
```

**Listing 3.2**
Command line interpreter.

Unlike Windows, file names in Linux are *case-sensitive*. `Foobar` is a different file from `foobar` is different from `fooBar`. Sorting is also case-sensitive. File names beginning with upper case letters appear before those that begin with lower case letters in directory listings sorted by name[3]. File names that begin with "." are considered to be "hidden," and are not displayed in directory listings unless you specifically ask for them.

Additionally, the Linux file system has a number of features that go beyond what you find in a typical Windows system. Let us take a look at some of the features that may be of interest to embedded programmers.

### File Permissions

Because Linux is multiuser, every file has a set of *permissions* associated with it to specify what various classes of users are allowed to do with that file. Get a detailed listing of some

[3] The graphical file manager offers case insensitive sort as an option.

Linux directory, either by entering the command `ls –l` in a command shell window, or with the desktop file manager. Part of the entry for each file is a set of 10 flags and a pair of names that look something like this:

```
-rw-r--r-- Andy physics
```

In this example, Andy is the *owner* of the file, and the file belongs to a *group* of users called physics, perhaps the physics department at some university. Generally, but not always, the owner is the person who created the file.

The first of the 10 flags identifies the file type. Ordinary files get a dash here. Directories are identified by "d," links are "l," and so on. We will see other entries for this flag when we get to device drivers later. The remaining nine flags divide into three groups of three flags each. The flags are the same for all groups, and represent respectively permission to read the file, "r," write the file, "w," or execute the file if it's an executable, "x." Write permission also allows the file to be deleted.

The three groups then represent the permissions granted to different classes of users. The first group identifies the permissions granted the owner of the file, and virtually always allows reading and writing. The second flag group gives permissions to other members of the same group of users. In this case, the physics group has read access to the file but not write access. The final flag group gives permissions to the "world," i.e., all users.

The "x" permission is worth a second look. In Windows, a file with the extension `.exe` is *assumed* to be executable. In Linux, a binary executable is identified by the "x" permission, since we don't have an explicit file extension to identify it. Furthermore, only those classes of users whose "x" permission is set are allowed to invoke execution of the file. So if I am logged in as an ordinary user, I am not allowed to invoke programs that might change the state of the overall system, such as changing the network address.

Another interesting feature of "x" is that it also applies to shell scripts, which we will come to later in this chapter. For you DOS fans, a shell script is the same thing as a `.bat` file. It is a text file of commands to be executed as a program. But the shell won't execute the script unless its "x" bit is set.

### The "Root" User

There is one very special user, named "root," in every Linux system. root can do anything to any file, regardless of the permission flags. Root is primarily intended for system administration purposes, and is not recommended for day-to-day use. Clearly, you can get into a lot of trouble if you are not careful, and root privileges pose a potential security threat. Nevertheless, the kinds of things that embedded and real-time developers do with the system often require write or executable access to files owned by root, and thus require you to be logged in as the root user.

In the past, I would just log in as root most of the time, because it was less hassle. One consequence of this is that every file I created was owned by root, and could not be written by an ordinary user without changing the permissions. It became a vicious circle. The more I logged in as root, the more I *had* to log in as root to do anything useful. I've since adopted the more prudent practice of logging in as a normal user, and only switching to root when necessary.

If you are logged on as a normal user, you can switch to being root with either the `su`, **s**ubstitute **u**ser, or `sudo` commands. The `su` command with no arguments starts up a shell with root privileges, provided you enter the correct root password. To return back to normal user status, terminate the shell by typing ^d or `exit`.

The `sudo` command allows you to execute a command as root, provided you are properly authorized to do so in the "sudoers file," `/etc/sudoers`. The sudoers file is, of course, owned by root, so only the root user can authorize sudoers. Once you have been authenticated by entering your own password, you may use `sudo` for a short period of time (default 5 minutes) without reentering your password.

For example, if I wanted to change the permissions on a file in the `/dev` directory, I could execute:

```
sudo chmod o + rw /dev/ttyS0
```

I would be prompted for my password and, if entered successfully, the command would be executed. I could then continue to execute `sudo` commands for 5 minutes without having to reenter my password. Note that `sudo` offers better security than `su`, because the root user must authorize sudoers and the root password does not need to be disclosed.

## The /Proc File System

The `/proc` file system is an interesting feature of Linux. It acts just like an ordinary file system. You can list the files in the `/proc` directory, you can read and write the files, but they don't really exist. The information in a `/proc` file is generated on the fly when the file is read. The kernel module that registered a given `/proc` file contains the functions that generate read data and accept write data.

The `/proc` files are a window into the kernel. They provide dynamic information about the state of the system in a way that is easily accessible to user-level tasks and the shell. In the abbreviated directory listing of Fig. 3.9, the directories with number labels represent processes. Each process gets a directory under `/proc`, with several directories and files describing the state of the process.

```
ls -l /proc
total 0
dr-xr-xr-x   3 root     root         0 Aug 25 15:23 1
dr-xr-xr-x   3 root     root         0 Aug 25 15:23 2
dr-xr-xr-x   3 root     root         0 Aug 25 15:23 3
dr-xr-xr-x   3 bin      root         0 Aug 25 15:23 303
dr-xr-xr-x   3 nobody   nobody       0 Aug 25 15:23 416
dr-xr-xr-x   3 daemon   daemon       0 Aug 25 15:23 434
dr-xr-xr-x   3 xfs      xfs          0 Aug 25 15:23 636
dr-xr-xr-x   4 root     root         0 Aug 25 15:23 bus
-r--r--r--   1 root     root         0 Aug 25 15:23 cmdline
-r--r--r--   1 root     root         0 Aug 25 15:23 cpuinfo
-r--r--r--   1 root     root         0 Aug 25 15:23 devices
-r--r--r--   1 root     root         0 Aug 25 15:23 filesystems
dr-xr-xr-x   2 root     root         0 Aug 25 15:23 fs
dr-xr-xr-x   4 root     root         0 Aug 25 15:23 ide
-r--r--r--   1 root     root         0 Aug 25 15:23 interrupts
-r--r--r--   1 root     root         0 Aug 25 15:23 ioports
```

**Figure 3.9**
The /proc file system.

### Try it out

It is interesting to see how many processes Linux spawns just by booting up. Reboot your system, open a command shell and execute:

```
ps –A | wc
```

The `ps` command lists the processes running on the system. The output of ps is one line per process. The `wc` command counts the number of lines, words, and characters passed to it. The number of lines is essentially the number of processes running. On my CentOS 7 system, there are 188 processes running.

Now try:

```
ps –A | more
```

This command lets you see the output of `ps` one page at a time. Note that the information for the `ps` command comes from the `/proc` file system.

Here's another command that highlights the real-time nature of `/proc` data. Execute:

```
cat /proc/interrupts
```

The `interrupts` file lists all of the interrupt sources in the system, the number of times each interrupt has fired off since the system was booted, and the driver registered to the interrupt. Now execute the command again:

```
cat /proc/interrupts
```

You will see that many of the numbers have gone up, thus proving that the data is being generated dynamically.

## The Filesystem Hierarchy Standard

A Linux system typically contains a very large number of files. For example, a typical CentOS installation may contain upwards of 30,000 files occupying several GB of disk space. Clearly it is imperative that these files be organized in some consistent, coherent manner. That is the motivation behind the Filesystem Hierarchy Standard (FHS). The standard allows both users and software developers to "predict the location of installed files and directories"[4]. FHS is by no means specific to Linux. It applies to Unix-like operating systems in general.

The directory structure of a Linux file system always begins at the *root*, identified as "/." FHS specifies several directories and their contents directly subordinate to the root. This is illustrated in Fig. 3.10. The FHS starts by characterizing files along two independent axes:

- Sharable versus nonsharable. A networked system may be able to mount certain directories through Network File System, such that multiple users can share executables. On the other hand, some information is unique to a specific computer, and is thus not sharable.

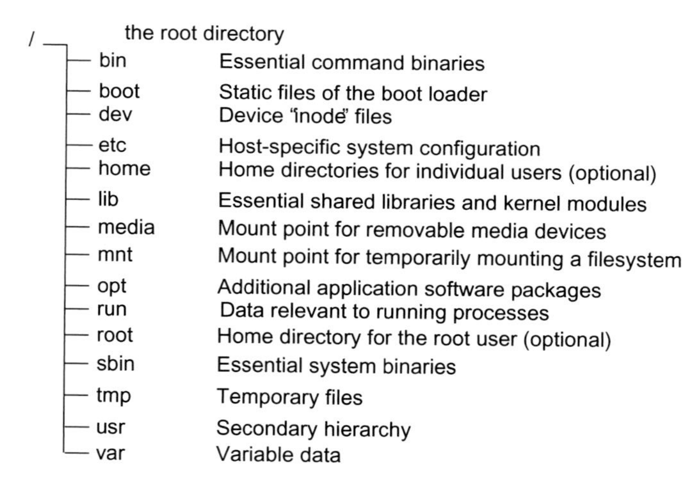


**Figure 3.10**
File system hierarchy.

[4] *Filesystem Hierarchy Standard—Version 3.0 dated 06/03/15*, maintained by the Linux Foundation. Available from https://wiki.linuxfoundation.org/en/FHS.

- Static versus. variable. Many of the files in a Linux system are executables that do not change, they are *static*. But the files that users create or acquire, by downloading or e-mail for example, are *variable*. These two classes of files should be cleanly separated.

Here is a description of the directories defined by FHS:

- `/bin` Contains binary executables of commands used both by users and the system administrator. FHS specifies what files `/bin` must contain. These include, among other things, the command shell and basic file utilities. `/bin` files are static and sharable.
- `/boot` Contains everything required for the boot process except configuration files and the map installer. In addition to the kernel executable image, `/boot` contains data that is used before the kernel begins executing user-mode programs. `/boot` files are static and nonsharable.
- `/etc` Contains host-specific configuration files and directories. With the exception of `mtab`, which contains dynamic information about file systems, `/etc` files are static. FHS identifies three optional subdirectories of `/etc`:
  - `/opt` Configuration files for add-on application packages contained in /opt.
  - `/sgml` Configuration files for SGML and XML
  - `/X11` Configuration files for X windows.

  In practice, most Linux distributions have many more subdirectories of `/etc` representing optional startup and configuration requirements.
- `/home` (Optional) Contains user home directories. Each user has a subdirectory under `home`, with the same name as his/her user name. Although FHS calls this optional, in fact it is almost universal among Unix systems. The contents of subdirectories under `/home` is, of course, variable.
- `/lib` Contains those shared library images needed to boot the system and run the commands in the root file system, i.e., the binaries in `/bin` and `/sbin`. In Linux systems `/lib` has a subdirectory, `/modules`, that contains kernel loadable modules.
- `/media` Mount point for removable media. When a removable medium is auto-mounted, the mount point is usually the name of the volume.
- `/mnt` Provides a convenient place to temporarily mount a file system.
- `/opt` Contains optional add-in software packages. Each package has its own subdirectory under `/opt`.
- `/run` Data relevant to running processes.
- `/root` Home directory for the root user[5]. This is not a requirement of FHS, but is normally accepted practice and highly recommended.

[5] Note the potential for confusion here. The directory hierarchy has both a *root*, "/", and a *root directory*, "/root", the home directory for the *root user*.

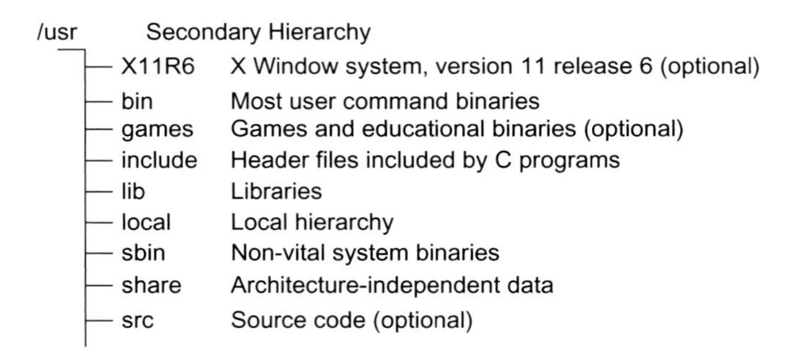


**Figure 3.11**
/usr hierarchy.

- /sbin Contains binaries of utilities essential for system administration such as booting, recovering, restoring, or repairing the system. These utilities are normally only used by the system administrator, and normal users should not need /sbin in their path.
- /tmp Temporary files.
- /usr Secondary hierarchy, see below.
- /var Variable data. Includes spool directories and files, administrative and logging data, and transient and temporary files. Basically, system-wide data that changes during the course of operation. There are a number of subdirectories under /var.

### *The /usr hierarchy*

/usr is a secondary hierarchy that contains user-oriented files. Fig. 3.11 shows the subdirectories under /usr. Several of these subdirectories mirror functionality at the root. Perhaps the most interesting subdirectory of /usr is /src for source code. This is where the Linux source is generally installed. You may in fact have sources for several Linux kernels installed in /src under subdirectories with names of the form:

```
linux-<version number>-ext
```

You would then have a logical link named linux pointing to the kernel version you are currently working with.

## *"Mounting" File Systems*

A major difference between Windows and Linux file systems has to do with how file structured devices, hard disks, floppy drives, CDROMs, etc., are mapped into the system's directory or hierarchy structure. The Windows file system makes devices explicitly visible, identifying them with a letter-colon combination, as in "C:." Linux, on the other hand, emphasizes a *unified* file system in which physical devices are effectively rendered invisible.

The mechanism that maps physical devices into the file system's directory structure is called "mounting"[6]. Removable media devices such as the CD-ROM drive are the most visible manifestation of this feature. Before you can read a CD-ROM, you must mount the drive onto an existing node in the directory structure using the `mount` command as in:

```
mount /media/cdrom
```

This command works because the file `/etc/fstab` has information about what device is normally mounted at `/media/cdrom`, and what type of file system, in this case iso9660, the device contains.

Like file permissions, `mount` can sometimes be a nuisance if all you want to do is read a couple files off a CD. But the real value of the mount paradigm is that it is not limited to physical devices directly attached to the computer, nor does it only understand native Linux file systems. As we will see later, we can mount parts of file systems on remote computers attached to a network to make their files accessible on the local machine. It is also possible to mount a device that contains a DOS FAT or VFAT file system. This is particularly useful if you build a "dual-boot" system that can boot up into either Windows or Linux. Files can be easily shared between the two systems. Or, if your embedded device has an SD or micro SD slot, or even USB, the ability to read VFAT file systems is extremely useful.

## System Configuration

The section above on the FHS mentioned the `/etc` directory. Here is one place where Unix systems really shine relative to Windows. OK, there may be any number of ways that Unix outshines Windows. In any case, the `/etc` directory contains essentially all of the configuration information required by the kernel and applications in the form of plain text files. The syntax and semantics of these files is not always immediately obvious, but at least you can read them. The format of many of the kernel's configuration files is documented in man pages (see section below on help).

By contrast, Windows systems have made a point of hiding configuration information in a magical place called "the Registry" (cue ominous music). Mortal users are often cautioned to stay away from the Registry, because improper changes can render the system unbootable. In fact, the registry can only be changed by a special program, `regedit`, and it was only a few years ago that I finally figured out where the registry physically resides.

If you have ever mustered up the courage to run `regedit` and actually look at the registry, you have no doubt noticed that the entries are rather obscure symbols, and the values are even more obscure numeric values. I submit that this was absolutely intentional on

[6] The term probably harks back to the days of reel-to-reel tape drives onto which a reel of tape had to be physically "mounted."

Microsoft's part. They didn't want idiot consumers mucking around with their computers, and possibly bricking[7] them. Linux users, on the other hand, know what they are doing (it says here), and are entitled to see and modify configuration data to their heart's content.

## The Shell

One of the last things that happens as a Linux system boots up is to invoke the command interpreter program known as the *shell*. Its primary job is to parse commands you enter at the console, and execute the corresponding program. But the shell is much more than just a simple command interpreter. It incorporates a powerful, expressive interpretive programming language of its own. Through a combination of the shell script language and existing utility programs, it is quite possible to produce very sophisticated applications without ever writing a line of C code. In fact, this is the general philosophy of Unix programming. Start with a set of simple utility programs that do one thing and one thing well, and then link them together through the shell scripting language.

The shell scripting language contains the usual set of programming constructs for looping, testing, functions, and so on. But perhaps the neatest trick in the shell bag is the concept of "pipes." This is a mechanism for streaming data from one program to another. The metaphor is perfect. One program dumps bytes into one end of the pipe, while a second program takes the bytes out the other end.

Most Linux/Unix utility programs accept their input from a default device called "stdin." Likewise, they write output to a device called "stdout." Any error messages are written to "stderr." Normally, `stdin` is the keyboard, while `stdout` and `stderr` are the display. But we can just as easily think of `stdin` and `stdout` as two ends of a pipe.

`stdin` and `stdout` can be *redirected*, so that we can send output to, for example, a file or take input from a file. In a shell window, try typing just the command `cat` with no arguments. `cat`[8] is perhaps the simplest of all Unix utilities. All it does is copy `stdin` to `stdout`, line by line. When you enter the command with no arguments, it patiently waits at `stdin` for keyboard input. Enter a line of text, and it will send that line to `stdout`, the display. Type Crtl-C to exit the program.

Now try this:

```
cat>textfile
```

[7] "Bricking" refers to messing up the configuration of a computer so badly that it no longer boots, thus rendering it the equivalent of a brick. The term is used a lot in the embedded space. A more apt term for desktop boxes might be "boat anchoring."

[8] Short for con**cat**enate. Don't you just love Unix command names?

Enter a line of text. This time you do not see the line repeated because the " > " operator has *redirected* stdout to the file textfile. Type a few lines, then Ctrl-C to exit.

The final step in this exercise is:

```
cat<textfile
```

Voila! The file you created with the previous command shows up on the screen because the " < " operator redirected stdin to textfile. cat actually implements a shortcut, so that if you enter a filename as a command line argument, without the < operator, it takes that as an input file. That is:

```
cat textfile
```

The real power of pipes though is the "|" operator, which takes stdout of one program and feeds it to stdin of another program. When I did the above exercises, I created a textfile containing:

```
this is a file
of text from the keyboard
```

Now if I execute:

```
cat textfile | grep this
```

I get:

```
this is a file
```

grep, as you may have guessed, is a filter. It attempts to match its command line arguments against the input stream stdin. Lines containing the argument text are passed to stdout. Other lines are simply ignored. What happened here is that the output of cat became the input to grep.

In typical Unix fashion, grep stands for Get Regular Expression "something." I forget what the "p" stands for. Regular expressions are in fact a powerful mechanism, a scripting language if you will, for searching text. grep embodies the syntax of regular expressions.

Shell scripts and makefiles make extensive use of redirection and piping.

Some other shell features are worth a brief mention, because they can save a lot of typing. The shell maintains a command history that you can access with the up arrow key. This allows you to repeat any previous command, or edit it slightly to create a new, similar command. There is a history command that dumps out the accumulated command history of the system or some subset. The history is maintained in the file .bash_history in your home directory.

```
981  make clean
982  make
983  ls /var/lock
984  ls /var/lock/lockdev/
985  ls /var/lock/subsys/
986  minicom -w3
987  minicom -w
988  ll /dev/ttyUSB0
989  ipconfig
990  ifconfig
991  minicom -w
992  man sudo
993  man chmod
994  man wc
995  ps-A | wc
996  cat interrupts
997  cat /proc/interrupts
998  history
999  man history
1000  985
1001  history
1002  ifconfig
1003  clear
1004  history
[doug@dougs ~]$
```

**Figure 3.12**
history command.

Fig. 3.12 shows the last few lines of the history command on my system. To re-execute any command in the list, simply enter "!" followed by the command number in the history list as in:

```
!998
```

to re-execute history.

Another really cool feature is auto-completion, which attempts to complete filename arguments for you. Say I wanted to execute:

```
cat verylongtextfilename
```

I could try entering:

```
cat verylong<tab>
```

Provided the remainder of the filename is unique in the current directory, the shell will automatically complete it, saving me the trouble of typing the whole filename. The shell beeps if it cannot find a unique match. Then you just have to type a little more until the remainder is unique.

Finally, the "~" character represents your home directory. So from anywhere in the file system you can execute cd ~ to go to your home directory.

There are, in fact, several shell programs in common use. They all serve the same basic purpose, yet differ in details of syntax and features. The most popular are:

- Bourne Again SHell – `bash`. `bash` is a "reincarnation" of the Bourne shell, `sh`, originally written by Stephen Bourne and distributed with Unix version 7. It is the default on most Linux distributions, and you should probably use it unless you have good reason, or strong feelings, for using a different shell.
- Korn Shell – `ksh`. Developed by David Korn at AT&T Bell Laboratories in the early 1980s. It features more advanced programming facilities than `bash`, but nevertheless maintains backward compatibility.
- Tenex C Shell – `tcsh`. A successor to the C shell, `csh` that was itself a predecessor to the Bourne shell. Tenex was an operating system that inspired some of the features of `tcsh`.
- Z Shell – `zsh`. Described as an extended Bourne shell with a large number of improvements, including some of the most useful features of `bash`, `ksh`, and `tcsh`.

The subject of shell programming is worthy of a book in itself, and there are many. When I searched Amazon.com for "linux shell programming," I got 248 hits.

## Getting Help

The official documentation for a Unix/Linux system is a set of files called "man pages," man being short for manual. man pages are accessed with the shell command `man` as in:

```
man cp
```

to get help on the shell copy command. Try it. Type `man man` at the shell prompt to learn more about the `man` command. man presents the page a screen at a time with a ":" prompt on the bottom line. To move to the next screen, type <space>. To exit `man` before reaching the end of the page, type "q." You can also page up and page down through the man page. Interestingly enough, you will not find that information in the `man man` page. The writing style in man pages is rather terse; they're reference in nature, not tutorial. The information is typically limited to what the command does, what its arguments are, and what options it takes.

To make it easier to find a specific topic, the man pages are organized into sections as follows:

Section 1: User Commands entered at the shell prompt.
Section 2: The kernel API functions.
Section 3: C library functions.
Section 4: Devices. Information on specific peripheral devices.

Section 5: File formats. Describes the syntax and semantics for many of the files in `/etc`.
Section 6: Games.
Section 7: Miscellaneous
Section 8: System Administration. Shell commands primarily used by the system administrator.

Another useful source of information is "info pages." Info pages tend to be more verbose, providing detailed information on a topic. Info pages are accessed with the info command. Try this one:

```
info gcc
```

to learn more about the GCC compiler package.

In the graphical desktop environments, KDE and GNOME, you can also access the man and info pages graphically. I find this especially useful with the info pages to find out what is there.

Finally, no discussion of getting help for Linux would be complete without mentioning Google. When you are puzzling over some strange behavior in Linux, Google is your friend. One of my common frustrations is error messages, because they rarely give you any insight into what really went wrong. So type a part of the error message into the Google search box. You'll likely get back at least a dozen hits of forum posts that deal with the error, and chances are something there will be useful.

With a good basic understanding of Linux, our next task is to configure the development workstation and install the software that will allow us to develop target applications.

## Resources

Sobell, Mark G., *A Practical Guide to Linux*. This book has been my bible and constant companion since I started climbing that steep Linux learning curve. It is an excellent beginner's guide to Linux and Unix-like systems in general, although having been published in 1997 it is getting a bit dated and hard to find. It has been superseded by...

Sobell, Mark G., *A Practical Guide to Linux Commands, Editors, and Shell Programming*

tldp.org – The Linux Documentation Project. As the name implies, this is the source for documentation on Linux. You will find how-to's, in-depth guides, FAQs, man pages, even an on-line magazine, the *Linux Gazette*. The material is generally well-written and useful.

CHAPTER 4

# The host development environment

## Chapter Outline



*When you say "I wrote a program that crashed Windows", people just stare at you blankly and say "Hey, I got those with the system,* ***for free****".*

***Linus Torvalds***

In this chapter we will configure our host workstation environment in preparation for configuring the target board in the next chapter. The steps involved include:

- Installing some software from an ISO image that you will download from the Elsevier website. This includes:
  - Cross-tool chain;
  - Some sample code;
- Configuring the workstation;
- Configuring networking on the workstation.

We have got a lot to do in this chapter. But before we dive into that, let us take a look at the cross-development environment in general.

Linux for Embedded and Real-time Applications.
DOI: http://dx.doi.org/10.1016/B978-0-12-811277-9.00004-3

## Cross-Development Tools: the GNU Tool Chain

Not only is the target computer limited in resources, it may be a totally different processor architecture from your (probably) x86-based development workstation. Thus, we need a cross-development tool chain that runs on the PC but may have to generate code for a different processor. We also need a tool that will help us debug code running on that (possibly different) target.

### Gnu Compiler Collection

By far the most widely used compiler in the Linux world is GCC, the Gnu Compiler Collection. It was originally called the Gnu C Compiler, but the name was changed to reflect its support for more than just the C language. GCC has language front ends for C, C++, Objective C, Fortran, Java, and Ada, as well as run-time libraries for these languages.

GCC also supports a wide range of target architectures in addition to the x86. Supported targets include:

- Alpha
- ARM
- M68000
- MIPS
- PowerPC
- SPARC

among others. In fact, the Wikipedia article on GCC lists 43 processors and processor families supported by the Free Software Foundation version of GCC.

GCC can run in a number of operating environments, including Linux and other variants of Unix. There are even versions that run under DOS and Windows.

GCC can be built to run on one architecture (a PC for example), while generating code for a different architecture (an ARM perhaps). This is the essence of cross-development.

GCC is an amalgam of compiler, assembler, and linker. Individual language front ends translate source code into an intermediate assembly language format that is then assembled by a processor-specific assembler, and ultimately linked into an executable. For a program consisting of a single C source file, the command to build an executable can be as simple as:

```
gcc myprogram.c
```

By default this creates an executable called `a.out`. The –o option lets you do the obvious thing and rename the executable thus:

```
gcc –o myprogram myprogram.c
```

If your project consists of multiple source files, the −c option tells GCC to compile only.

```
gcc –c myprogram.c
```

generates the relocatable object file `myprogram.o`. Later this can be linked with other .o files with another GCC command to produce an executable.

Generally though, you will not be invoking GCC directly. You will use a *makefile* instead.

## Make

Real-world projects consist of anywhere from dozens to thousands of source files. These files have dependencies such that if you change one file, a header for example, another file, a C source, needs to be recompiled, and ultimately the whole project needs to be rebuilt. Keeping track of these dependencies, and determining what files need to be rebuilt at any given time, is the job of a powerful tool called the Make utility.

In response to the shell command `make`, the Make utility takes its instructions from a file named `Makefile` or `makefile` in the current directory. This file consists of *dependency lines* that specify how a *target file* depends on one or more *prerequisite files*. If any of the prerequisite files are more recent than the target, `make` updates the target using one or more *construction commands* that follow the dependency line. The basic syntax is:

```
target: prerequisite-list
<tab> construction commands
```

One common mistake to watch out for when writing makefiles is substituting spaces for the leading tab in the construction commands. It *must* be a tab.

Do not let this simple example fool you. There is a great deal of complexity and subtlety in the `make` syntax, which looks something like the shell scripting language. The sample code that you will download later in the chapter has some simple examples of makefiles. In a later chapter, you will configure and build the Linux kernel, which has a huge, very involved makefile.

Often the best way to create a makefile for a new project is to start with the one for an existing, similar project.

## Gnu DeBugger

GDB stands for the "Gnu DeBugger." This is a powerful source-level debugging package that lets you see what is going on inside your program. You can step through the code, set breakpoints, examine and change variables, and so on. Like most Linux tools, GDB itself is command line driven, making it rather tedious to use. There are several graphical front ends for GDB that translate GUI commands into GDB text commands. Eclipse, which you will encounter in the next chapter, has an excellent graphical front end for GDB.

GDB can be set up to run on a host workstation while debugging code on a separate target machine. The two machines can be connected via serial or network ports, or you can use an in-circuit emulator (ICE). Many ICE vendors have GDB back ends for their products.

There are also specialized debug ports built into some processor chips. These go by the names Joint Test Action Group and Background Debug Mode. These ports provide access to debug features such as hardware breakpoint registers. GDB back ends exist for these ports as well.

## Getting and Installing the Software

The recommended target board for the examples in the book is the BeagleBone Black, an open source, or as the website puts it, "community supported" project. The resources section has links for ordering and other resources. Chapter 6, The hardware contains more details on the board.

Throughout the book we will be installing software from different web sources. For now, navigate to the book's home page at the Elsevier web site, as listed in the Resources section. Download the file `EmbeddedLinux4.iso`. A `.iso` file is an image of an optical disk, either a CD or DVD. The name derives from ISO[1] 9660, which is the standard for optical disk file systems.

The file probably ended up in `$HOME/Downloads`. To access the `.iso` file, you will need to mount it as if it were in fact a disk.

- As root user create a directory under `/mnt` named `iso/`.
- `mount –t iso9660 $HOME/Downloads/EmbeddedLinux4.iso /mnt/iso`.

### Install Cross-Tool Chain

Cross-tool chains are available on the web. In particular, the Linaro project offers version 6.3 of an ARM GCC tool chain. But there is a problem. The kernel will not build in a version 6.x compiler. The recommended version is 4.9. I could not find any prebuilt copy of gcc 4.9, so I was forced to build my own using a tool named Crosstools-NG. I will save you the trouble of building the tool chain, but if you are interested, visit the Crosstools-NG website listed in the Resources section.

[1] International Standards Organization.

As root user copy /mnt/iso/arm-gcc-4.9.4.tar.bz2 to /usr/local and untar it like this:

```
cp /mnt/iso/arm-gcc-4.9.4.tar.bz2 /usr/local
cd /usr/local
tar –xf arm-gcc-4.9.4.tar.bz2
```

In the course of the book, you will be doing a lot of "untarring," the third command in the list above. So, when you see instructions to untar a certain file, that is the command you will use.

The tool chain is installed in arm-unknown-linux-gnueabi/. You will need to add /usr/local/arm-unknown-linux-gnueabi/bin to your PATH, the list of directories that the shell searches to find executables. In CentOS, the file that defines the PATH is .bash_profile. In Ubuntu/Debian installations, it is just .profile. From here on, we will be doing a lot of editing. If you have a favorite Linux editor, by all means use it. Personally, I refuse to touch vi or emacs[2]. My favorite K Desktop Environment (KDE) editor is Kwrite. Double-click a text file in a file manager window, and it automatically opens in Kwrite.

In .bash_profile (or .profile) you will find a line that begins PATH=$PATH:, probably followed by $HOME/bin. At the end of that line add ":/usr/local/arm-unknown-linux-gnueabi/bin". Save the file and close the editor. The new path will not take effect until you log out and log back in again. No hurry. It will be a while yet before we get around to building a program.

## *The Terminal Emulator,* minicom

minicom is a fairly simple Linux application that emulates a dumb RS-232 terminal through a serial port. This is what we will use to communicate with the target board.

Recent Linux distributions tend not to install minicom by default. It is easy enough to install. As root user (again, either su or sudo) execute:

yum install minicom    in CentOS
apt-get install minicom    in Ubuntu or Debian

Both of these commands do essentially the same thing. They go out on the Internet to find the specified package, resolve any dependencies, download the packages, and install them.

Once minicom is installed, there are a number of configuration options that we need to change to facilitate communication with the target.

[2] Although I have had my arm twisted to try vim and it's not too bad.

In a shell window, again as *root user*, enter the command `minicom –s`. If you are running `minicom` for the first time, you may see the following warning message:

```
WARNING: Configuration file not found. Using defaults
```

You will be presented with a configuration menu. Select `Serial port setup` (Fig. 4.1). By default, `minicom` communicates through the modem device, `/dev/modem`. We need to change that to talk directly to a serial port. Our target board emulates a serial port on its USB interface. Type "A" and replace the word "`modem`" with "`ttyACM0`."

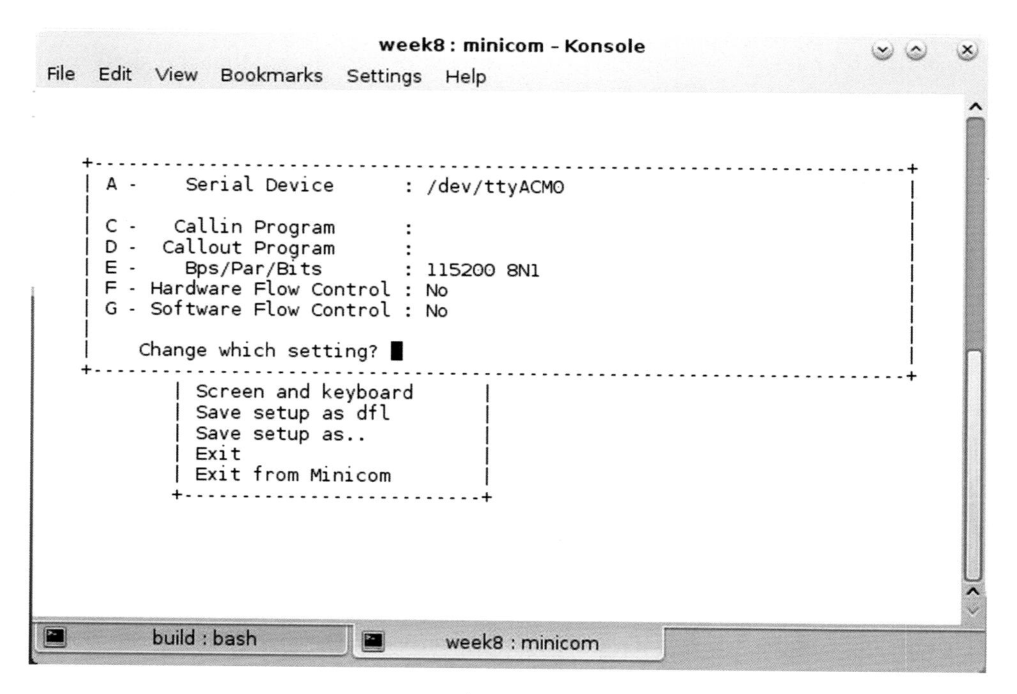


**Figure 4.1**
Minicom serial port setup menu.

Type "E" followed by "I" to select 115200 baud. Make sure both hardware and software flow control are set to "no." These are toggles. Typing "F" or "G" will toggle the corresponding value between "yes" and "no." Fig. 4.1 shows the final serial port configuration.

Type <Enter> to exit Serial port setup, and then select Modem and dialing. Here we want to delete the modem's Init string and Reset string, since they are just in the way on a direct serial connection. Type "A" and backspace through the entire Init string. Type "B," and do the same to the Reset string. The other strings in this menu do not matter, because they will never be invoked.

Type <Enter> to exit Modem and dialing, and then select Screen and keyboard. Type "B" once to change the Backspace key from BS to DEL.

Type <Enter> to exit Screen and keyboard. Select Save setup as dfl (default). Then select Exit.

At this point, `/dev/ttyACM0` does not exist because you have not powered up the target. But if it did exist, you would see that only the owner and group have access to the device. In order to gain access to this port, you must become a member of the `dialout` group.

There used to be a nice graphical dialog accessible from the Start menu under `Administration>Users and Groups` that made this process very easy. Not anymore. Now you have to edit the appropriate file under `/etc`., in this case `group`. As root user, open `/etc/group` in an editor. Find the line that begins "`dialout:`" and add "`:<your_user_name>`" at the end of the line. Save the file and close the editor.

## Networking

There are a couple of changes needed on the host's network configuration.

### Network Address

Your workstation is probably configured to get a network address via DHCP (Dynamic Host Configuration Protocol). But in this case, to keep things simple, we are going to specify fixed IP addresses for both the workstation and the target.

The CentOS graphical menu for changing network parameters is accessed from `Settings -> System Settings`[3]. Select `Network Settings` and either the `Wired` or `Wireless` tab, depending on the nature of your network connection. In the case of a virtual machine, even a wireless interface looks like a wired interface to the guest OS.

Select the network interface and click `Edit...` Your network interface is probably named something like `enp0s8`, which means it is an Ethernet device (en) attached to slot 8 of PCI bus segment 0. Change the Method from DHCP to Manual. Now enter the fixed IP address. Network address 192.168.1 is a good choice here, because it is one of a range of network addresses reserved for local networks. Select node 2 for your workstation. The Subnet Mask will be automatically filled in. Fig. 4.2 shows the final result. You may also want to set the Default Gateway Address and DNS nodes if you are connected to the Internet.

[3] Configuration options have an annoying tendency to move around with each new release of CentOS. This is where it is located in CentOS 7.

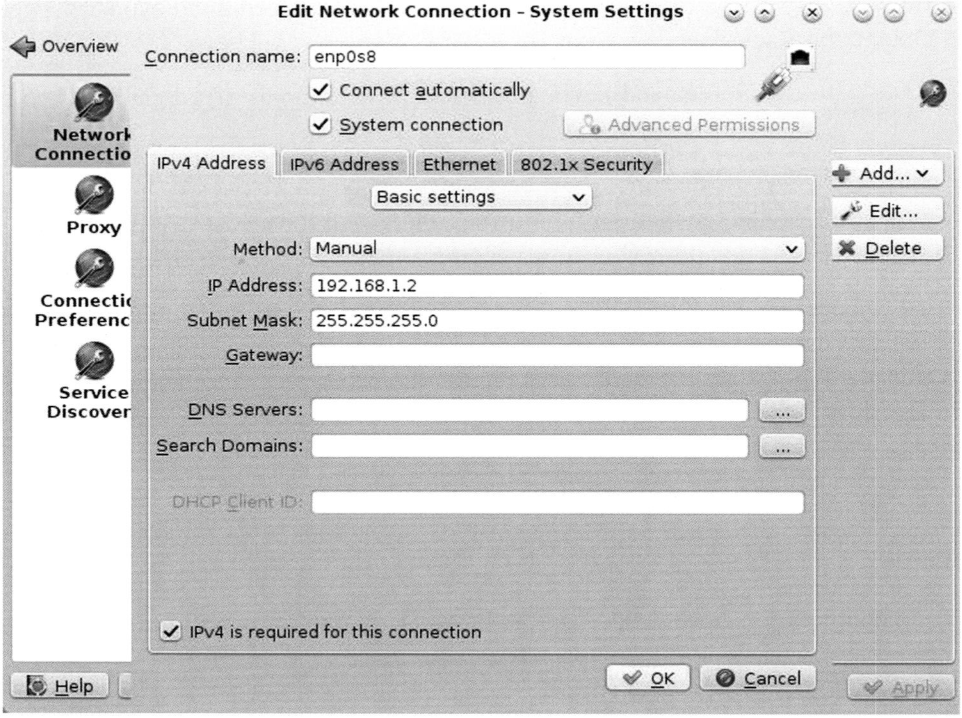

**Figure 4.2**
Edit network device dialog.

When you exit the `Edit Network Connections` dialog you are presented with a dialog about the KDE Wallet system asking you to choose an encryption method. To be honest, I've never understood what the KDE Wallet system is about, and have never used it, so I just click Cancel here.

Alternatively, you can just go in and directly edit the network device parameters file. Network configuration parameters are found in /`etc/sysconfig/network-scripts`/, where you should find a file named something like `ifcfg-enp0s8` that contains the parameters for network adapter enp0s8. You might want to make a copy of this file and name it something like `dhcp-ifcfg-enp0s8`. That way you will have a DHCP configuration file for future use if needed. Now, open the original file with an editor (as root user of course). It should look something like Listing 4.1A. The lines shown here may be in a different order, and interspersed with other lines.

Change the line `BOOTPROTO = dhcp` to `BOOTPROTO = none` and add the two new lines shown in Listing 4.1B. Remove the lines about PEERDNS and PEERROUTES.

| TYPE=Ethernet<br>BOOTPROTO=dhcp<br>DEFROUTE=yes<br>IPV4_FAILURE_FATAL=yes<br>IPV6INIT=no<br>NAME=enp0s8<br>UUID=27db13eb-6a1d-41e5-88bc-de83a6aa0895<br>ONBOOT=yes<br>PEERDNS=yes<br>PEERROUTES=yes | TYPE=Ethernet<br>BOOTPROTO=**none**<br>DEFROUTE=yes<br>IPV4_FAILURE_FATAL=yes<br>IPV6INIT=no<br>NAME=enp0s8<br>UUID=27db13eb-6a1d-41e5-88bc-de83a6aa0895<br>ONBOOT=yes<br>**IPADDR=192.168.1.2**<br>**PREFIX=24** |
|---|---|
| Listing 4-1a: ifcfg-eth0 | Listing 4-1b: revised ifcfg-eth0 |

**Listing 4.1**
(A) ifcfg-eth0; (B) revised ifcfg-eth0.

### *What About Wireless?*

Personally, I have not had any success configuring wireless network ports directly in Linux. But wireless ports do work fine in a virtual machine environment, because the VM manager virtualizes the network interface so that it looks like something else to the guest machine. VirtualBox, e.g., translates any physical network interface to an Intel 82540EM Gigabit interface.

Having two interfaces can be very useful. You could, e.g., leave the wireless port attached to your network getting its IP address through DHCP, and reconfigure the Ethernet port as described above for exclusive connection to the target. For example, my network is 192.168.1. I plug the target board into a switch. The DHCP server in the router serves node addresses ranging from 100 to 199. The Windows host OS on my laptop gets its IP address from DHCP. The Linux guest has its address fixed by the procedure described here, and everything gets along fine.

### *Network File System*

We use NFS (Network File System) to mount the target board's root file system[4]. The reason for doing that will become apparent when we start developing applications starting in Chapter 7, Accessing hardware. That means we have to "export" one or more directories on the workstation to make them visible on the network.

[4] Ubuntu distros tend to not have NFS installed by default. Execute `sudo apt-get install nfs-kernel-server` to install it.

Exported directories are specified in the file /etc/exports. The file probably exists, but is empty at this point. As root user, open it with an editor and add the following line:

```
/home/<your_home_name>*(rw,no_root_squash,sync,no_subtree_check)
```

Replace <your_home_name> with the name of your home directory. This makes your home directory visible on the network where other nodes can mount parts of it to their local file system. The "*" represents clients that are allowed to mount this directory. In this case, the directory is visible to any machine on the network. You can specify specific machines either by DNS name or IP address (as in machine1.intellimetrix.us, or 192.168.1.50, for example). You can also specify all the machines on a given network by giving the network and network mask as in 192.168.1.0/255.255.255.0.

The options inside the parentheses describe how the client machines can access the directory. Options shown here are:

- rw: Allow read and write access. Alternately, you can specify ro for read only.
- no_root_squash: By default, a file request made by the root user on a client machine is treated on the server as if it came from user nobody. If no_root_squash is specified, the root user on the client has the same level of access as the server's root user. This can have security implications, but is necessary in some circumstances.
- sync: Normally, file writing is asynchronous, meaning that the client will be informed that a write is complete when NFS hands the request over to the file system, but before it actually reaches its final destination. The sync option waits until the operation really is complete.
- no_subtree_check: If only part of a volume is exported, a routine called *subtree checking* verifies that a requested file is in the exported part. If the entire volume is exported, or you are not concerned about this check, disabling subtree checking will speed up transfers.

NFS is somewhat picky about the format of this file. The last line must end with an end-of-line character, and no spaces are allowed inside the parentheses.

With the exports file in place, we need to be sure that NFS is actually running. In CentOS, we do that with the systemctl shell command as shown in Fig. 4.3, which shows the result if the NFS server is running. The complete command is:

```
systemctl status nfs-server
```

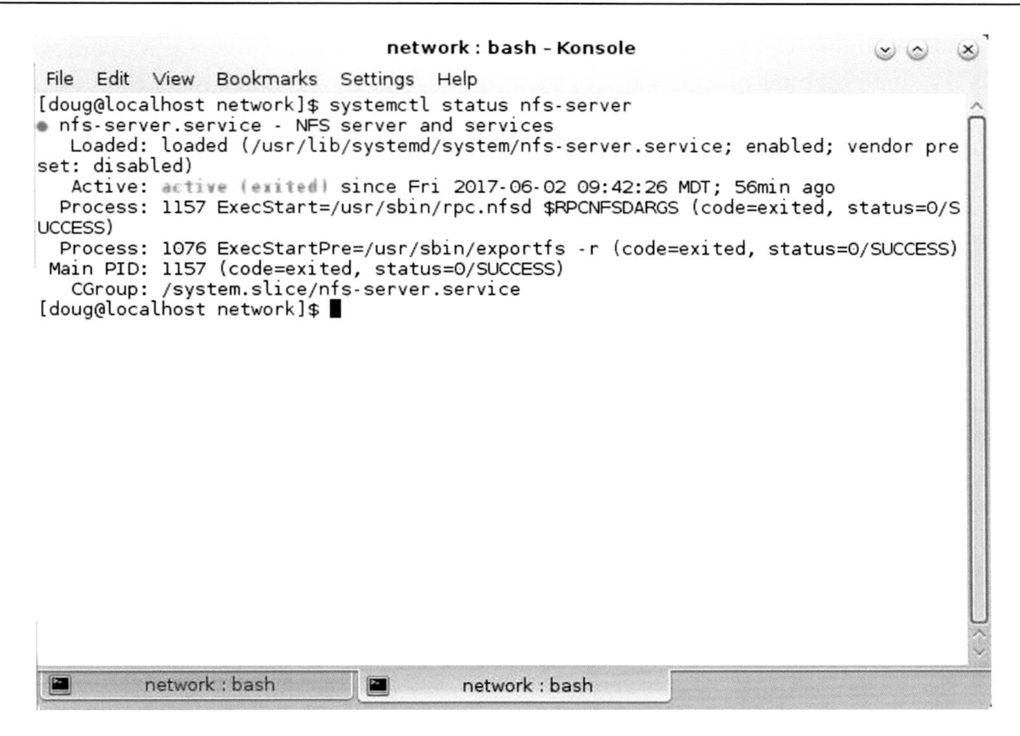


**Figure 4.3**
Systemctl command.

systemctl is the user interface to the systemd system and service manager. The arguments to systemctl are a command and the name of a service, which is represented by a script file in `/usr/lib/systemd/system`. If the NFS server is not running, you start it with this command:

```
systemctl start nfs-server
```

To stop it you would execute:

```
systemctl stop nfs-server
```

If, for example, you were to make changes to `/etc/exports`, you would want to restart NFS for the changes to take effect:

```
systemctl restart nfs-server
```

Look in `/usr/lib/systemd/system` to see that there is a script file named `nfs-server.service`. Have a look at the file to see how it works.

### *Trivial File Transfer Protocol*

The last item of business for this chapter is installing (if necessary) and enabling TFTP, the Trivial File Transfer Protocol. We will use that later when building a new Linux kernel.

As the name implies, TFTP is a simple protocol for file transfer that can be implemented in a relatively small amount of memory. As such, it is useful for things like booting devices that have no local mass storage. It uses UDP datagrams, and thus provides its own simplified session and transport mechanisms. No authentication or encryption mechanisms are provided, so it probably should not be used in situations where security is a concern.

The workstation will serve as our TFTP server, just as it provides the NFS server function. Some distributions do not install the TFTP server by default, and so you must install it yourself. Just as we did earlier with `minicom`, CentOS users can execute `yum install tftp-server`, while Ubuntu and Debian users can execute `apt-get install tftp-server`.

TFTP is one of several network protocols managed by `xinetd`, the eXtended InterNET Daemon. `xinetd` features access control mechanisms such as TCP Wrapper ACLs, extensive logging capabilities, and the ability to make services available based on time. It can place limits on the number of servers that the system can start, and has deployable defense mechanisms to protect against port scanners, among other things.

Once installed, TFTP needs to be enabled. We do that by editing the file `/etc/xinetd/tftp`. You will need to be root user to save any changes. Find the line that starts "disable =" and change "yes" to "no."

While you are in this file, also note the line that starts "server_args = ." The argument for this line is the default directory for TFTP transfers. Currently in CentOS it is `/var/lib/tftpboot`. Any files you want to download to the target board over TFTP need to be copied to this directory. But this directory is owned by root, and is not writable by the world. So you need to make it writable by everyone by executing, as root user:

```
chmod 777 /var/lib/tftpboot
```

We now have almost everything we need on the host workstation for embedded development. In the next chapter we will take a detour into the open source Integrated Development Environment Eclipse that will make application development a lot more productive. After that, we will set up the target hardware.

## Resources

https://www.elsevier.com/books-and-journals/book-companion/9780128112779 – Book's home page at Elsevier
crosstool-ng.org – website for the Crosstool-NG tool chain builder.
beagleboard.org – website for the whole family of BeagleBoards. Among other things, here are links to distributors of the BeagleBone Black.
Mecklenburg, Robert, *Managing Project with GNU Make 3rd Edition*, O'Reilly, 2005. This is the bible when you are ready to fully understand and exploit the power of `make`.

CHAPTER 5

# Eclipse integrated development environment[1]

## Chapter Outline



> *You could spend* all day *customizing the title bar. Believe me. I speak from experience.*
>
> ***Matt Welsh***

[1] Portions of this chapter are adapted from *Eclipse Platform Technical Overview*, © IBM Corporation and The Eclipse Foundation, 2001, 2003, 2006, and made available under the Eclipse Public License (EPL). The full document is available here: https://eclipse.org/articles/Whitepaper-Platform-3.1/eclipse-platform-whitepaper.html.


Linux for Embedded and Real-time Applications.
DOI: http://dx.doi.org/10.1016/B978-0-12-811277-9.00005-5

Integrated development environments (IDE) are a great tool for improving productivity. Desktop developers have been using them for years. Perhaps the most common example is Microsoft's Visual Studio environment. In the past, a number of embedded tool vendors have built their own proprietary IDEs.

In the Open Source world, the IDE of choice today is Eclipse, also known as the Eclipse Platform, and sometimes just the Platform. The project was started by Object Technology International, which was subsequently purchased by IBM. In late 2001, IBM and several partners formed an association to promote and develop Eclipse as an Open Source project. It is now maintained by the Eclipse Foundation, a nonprofit consortium of software industry vendors. Several leading embedded Linux vendors such as Monta Vista, TimeSys, LinuxWorks, and Wind River Systems have embraced Eclipse as the basis for their future IDE development.

I would go so far as to say that Eclipse is the most professional, well-run open source project out there. Every year for over 10 years, in the middle of June, the foundation publishes a major new release of the platform and most of the related projects, amounting to something like 20 to 30,000,000 lines of code. Up until 2011, these releases were named after the moons of Jupiter in alphabetical order. The 2010 release was called Helios. The 2011 release broke with that tradition and was called Indigo.

It should be noted that Eclipse is not confined to running under Linux. It runs just as well under various Windows operating systems.

## Overview

"Eclipse is a kind of universal tool platform—an open, extensible IDE for anything and nothing in particular. It provides a feature-rich development environment that allows the developer to efficiently create tools that integrate seamlessly into the Eclipse platform," according to the platform's own on-line overview. Technically, Eclipse itself is not an IDE, but is rather an *open platform* for developing IDEs and *rich client* applications.

Fig. 5.1 shows the basic Eclipse workbench. It consists of several *views* including:

- Navigator – shows the files in the user's workspace;
- Text Editor – shows the contents of a file;
- Tasks – a list of "to dos";
- Outline – of the file being edited. The contents of the outline view are content-specific.

Although Eclipse has a lot of built-in functionality, most of that functionality is very generic. It takes additional tools to extend the platform to work with new content types, to do new things with existing content types, and to focus the generic functionality on something specific.

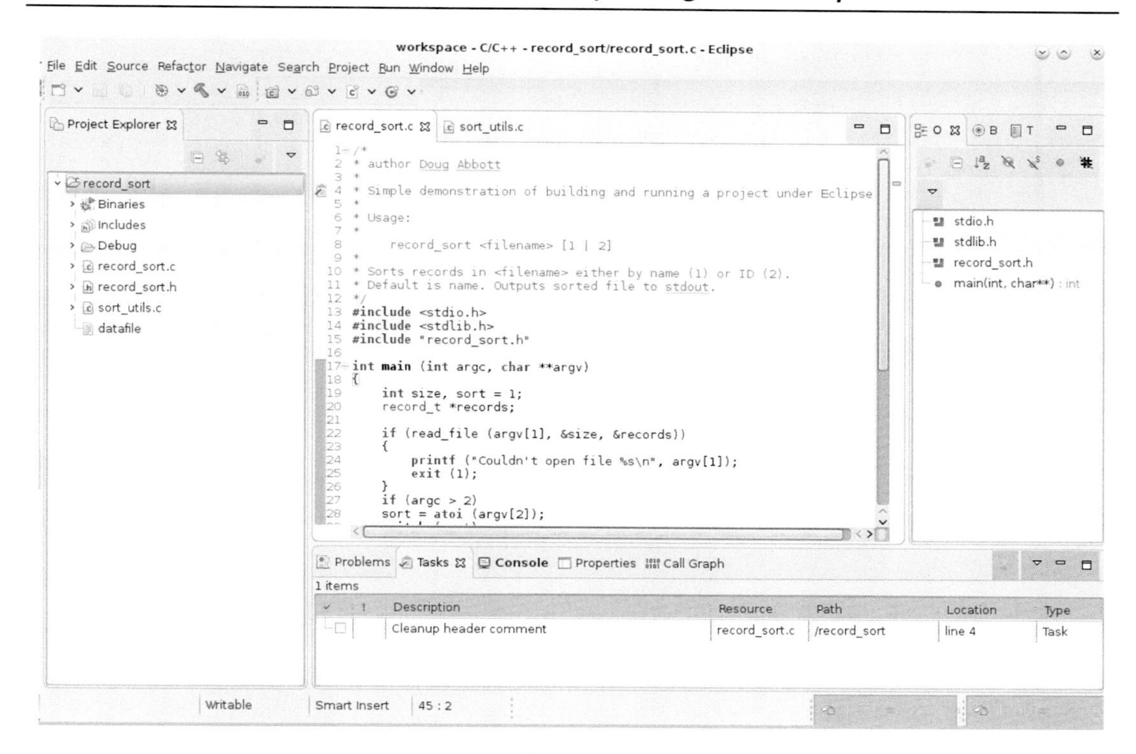


**Figure 5.1**
The Eclipse workbench.

The Eclipse Platform is built on a mechanism for discovering, integrating, and running modules called *plug-ins*. A tool provider writes a tool as a separate plug-in that operates on files in the workspace and exposes its tool-specific UI in the workbench. When you launch Eclipse, it presents an IDE composed of the set of available plug-ins.

The Eclipse platform is written primarily in Java, and in fact was originally designed as a Java development tool. Fig. 5.2 shows the platform's major components and Application Programming Interfaces (APIs). The platform's principal role is to provide tool developers with mechanisms to use, and rules to follow, for creating seamlessly integrated tools. These mechanisms are exposed via well-defined API interfaces, classes, and methods. The platform also provides useful building blocks and frameworks that facilitate developing new tools.

Eclipse is designed and built to meet the following requirements:

- Support the construction of a variety of tools for application development.
- Support an unrestricted set of tool providers, including independent software vendors.

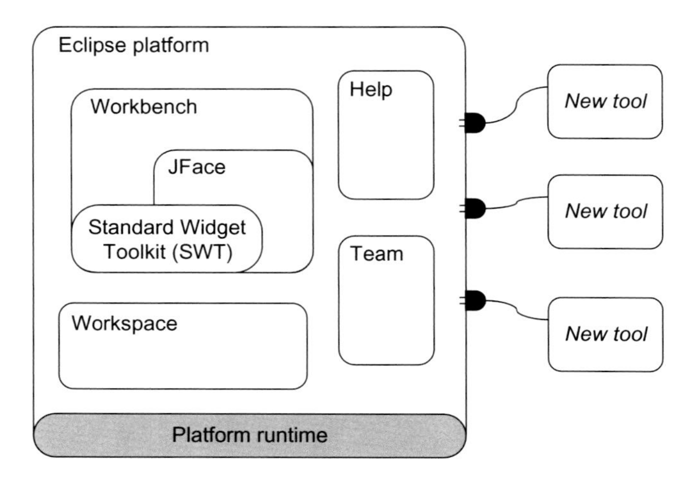


**Figure 5.2**
Eclipse platform.

- Support tools to manipulate arbitrary content types such as HTML, Java, C, JSP, EJB, XML, and GIF.
- Facilitate seamless integration of tools within and across different content types and tool providers.
- Support both GUI and non-GUI-based application development environments.
- Run on a wide range of operating systems, including Windows and Linux.
- Capitalize on the popularity of the Java programming language for writing tools.

## *Plug-ins*

A *plug-in* is the smallest unit of Eclipse functionality that can be developed and delivered separately. A small tool is usually written as a single plug-in, whereas a complex tool may have its functionality split across several plug-ins. Except for a small kernel known as the Platform Runtime, all of the Eclipse platform's functionality is located in plug-ins.

Plug-ins are coded in Java. A typical plug-in consists of Java code in a JAR library, some read-only files, and other resources such as images, web templates, message catalogs, native code libraries, etc. Some plug-ins do not contain code at all. An example is a plug-in that contributes on-line help in the form of HTML pages. A single plug-in's code libraries and read-only content are located together in a directory in the file system, or at a base URL on a server.

Each plug-in's configuration is described by a pair of files. The *manifest* file, `manifest.mf`, declares essential information about the plug-in, including the name, version, and dependencies to other plug-ins. The second optional file, `plugin.xml`, declares the plug-in's interconnections to other plug-ins. The interconnection model is simple: a plug-in declares any number of named *extension points*, and any number of *extensions* to one or more extension points in other plug-ins.

The extension points can be extended by other plug-ins. For example, the workbench plug-in declares an extension point for user preferences. Any plug-in can contribute its own user preferences by defining extensions to this extension point.

### *Workbench*

The Eclipse *workbench* API, user interface (UI), and implementation are built from two toolkits:

- Standard Widget Toolkit (SWT) – a widget set and graphics library integrated with the native window system, but with an OS-independent API.
- JFace – a UI toolkit implemented using SWT that simplifies common UI programming tasks.

Unlike SWT and JFace, which are both general purpose UI toolkits, the workbench provides the UI personality of the Eclipse Platform, and supplies the structures in which tools interact with the user. Because of this central and defining role, the workbench is synonymous with the Eclipse Platform UI as a whole, and with the main window you see when the Platform is running (see Fig. 5.1). The workbench API is dependent on the SWT API, and to a lesser extent on the JFace API.

The Eclipse Platform UI paradigm is based on editors, views, and perspectives. From the user's standpoint, a workbench window consists visually of views and editors. Perspectives manifest themselves in the selection and arrangements of editors and views visible on the screen.

*Editors* allow you to open, edit, and save objects. They follow an open-save-close lifecycle, much like file system-based tools, but are more tightly integrated into the workbench. When active, an editor can contribute actions to the workbench menus and tool bar. The platform provides a standard editor for text resources; other plug-ins supply more specific editors.

*Views* provide information about some object that you are working with. A view may assist an editor by providing information about the document being edited. For example, the standard content outline view shows a structured outline for the content of the active editor, if one is available. A view may augment other views by providing information about the currently selected object. For example, the standard properties view presents the properties of the object selected in another view. The platform provides several standard views; additional ones are supplied by other plug-ins.

A workbench window can have several separate *perspectives*, only one of which is visible at any given moment. Each perspective has its own views and editors that are arranged (tiled, stacked, or detached) for presentation on the screen, although some may be hidden. Several different types of views and editors can be open at the same time within a perspective. A perspective controls initial view visibility, layout, and action visibility. You can quickly switch perspectives to work on a different task, and can easily rearrange and customize a perspective to better suit a particular task. The platform provides standard perspectives for general resource navigation, on-line help, and team support tasks. Other plug-ins provide additional perspectives.

Plug-in tools may augment existing editors, views, and perspectives to:

- Add new actions to an existing view's local menu and tool bar.
- Add new actions to the workbench menu and tool bar when an existing editor becomes active.
- Add new actions to the pop-up content menu of an existing view or editor.
- Add new views, action sets, and shortcuts to an existing perspective.

## Installation

There are two parts to Eclipse; Eclipse itself and the Java Runtime Engine (JRE). Download Eclipse from `eclipse.org`. There are several versions to choose from. Since we are interested in C development, get the Eclipse IDE for C/C++ Developers.

You can put Eclipse wherever you like. I chose to install it in `/usr/local`. After untarring, you will find a new subdirectory called, not surprisingly, `eclipse/`.

Since Eclipse is written primarily in Java, you'll need a JRE to support it. Most contemporary Linux distributions install a JRE by default, but it may not be compatible with Eclipse. The version supplied with CentOS 7 works fine. If necessary, download the latest version from Oracle (formerly Sun Microsystems). The download is available either as a binary executable or an RPM package. Copy it wherever you like and execute it. `/usr/local` is a good place. You will be asked to read and accept the Binary Code License Agreement for the Java SE runtime environment (JRE) version 6. You will need to add the`bin/` directory of the newly-installed JRE to your PATH.

That's it! Eclipse is installed and ready to go. In a shell window, `cd` to the `eclipse/` directory and execute `./eclipse`. From anywhere else, enter the complete path to the eclipse executable. Or just double-click it in a K Desktop Environment file manager window.

Or better yet, create a script file named `eclipse` in directory `/usr/bin`. It contains the following:

```
#!/bin/bash
/usr/local/eclipse/eclipse &
```

The first line identifies this file as a shell script. The second line executes Eclipse. The "&" says run this program in the background and return the shell prompt. Since `/usr/bin` is in your path, all you have to type is `eclipse`.

## Using Eclipse

Every time Eclipse starts up, it asks you to select a *workspace*, a place where your project files are held. The default is a directory called `workspace/` under your home directory. Under `workspace/` then are subdirectories for each project. Following the workspace dialog, if this is the first time you have executed Eclipse, you'll see the Welcome screen. On subsequent runs, Eclipse opens directly in the workbench, but you can always access this window from the first item on the Help menu, Welcome.

The Tutorial icon on the Welcome screen leads you to a very good, very thorough introduction to using Eclipse. If you are new to the Eclipse environment, I strongly suggest you go through it. There is no point in duplicating it here. Go ahead, I'll wait.

OK, now you have a pretty good understanding of generic Eclipse operation. It is time to get a little more specific, and apply what we have learned to C software development for embedded systems. Since you are running the CDT version of Eclipse, you will see the C/C++ perspective initially.

## The C Development Environment (CDT)

Among the ongoing projects at Eclipse.org is the tools project. A subproject of that is the C/C++ Development Tools (CDT) project that has built a fully functional C and C++ IDE for the Eclipse platform. CDT under Linux is based on GNU tools and includes:

- C/C++ Editor with syntax coloring;
- C/C++ Debugger using GDB;
- C/C++ Launcher for external applications;
- Parser;
- Search Engine;
- Content Assist Provider;
- Makefile generator.

### *Creating a New Project*

For our initial exercise in creating a C/C++ project, we will create a fairly straightforward record sorting application. The records to be sorted consist of a name and an ID number. To simplify the code a bit, we will replace any spaces in the name field with underscores. The program will sort a file of records in ascending order, either by name or ID number, as specified on the command line thus:

```
record_sort <datafile> [1 | 2]
```

where 1 means sort by name, and 2 means sort by ID. Sort by name is the default if no sorting argument is given.

In the Eclipse C/C++ perspective, select `File>New>C Project`. This brings up the New Project wizard shown in Fig. 5.4. Call it "record_sort." The project type is `Executable> Empty Project`, The tool chain is `Linux GCC`, and we will use the default workspace location. Clicking `Next` brings up the Select Configurations dialog, where you can select either or

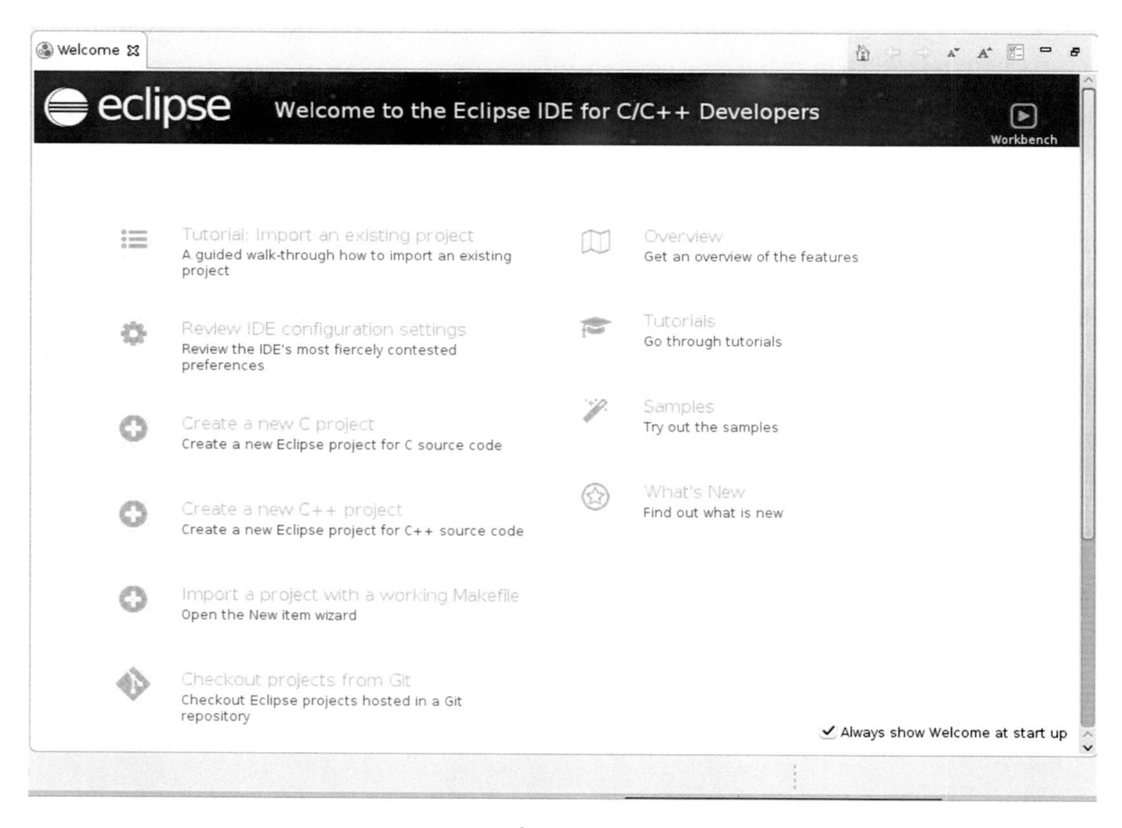


**Figure 5.3**
Eclipse Welcome screen.

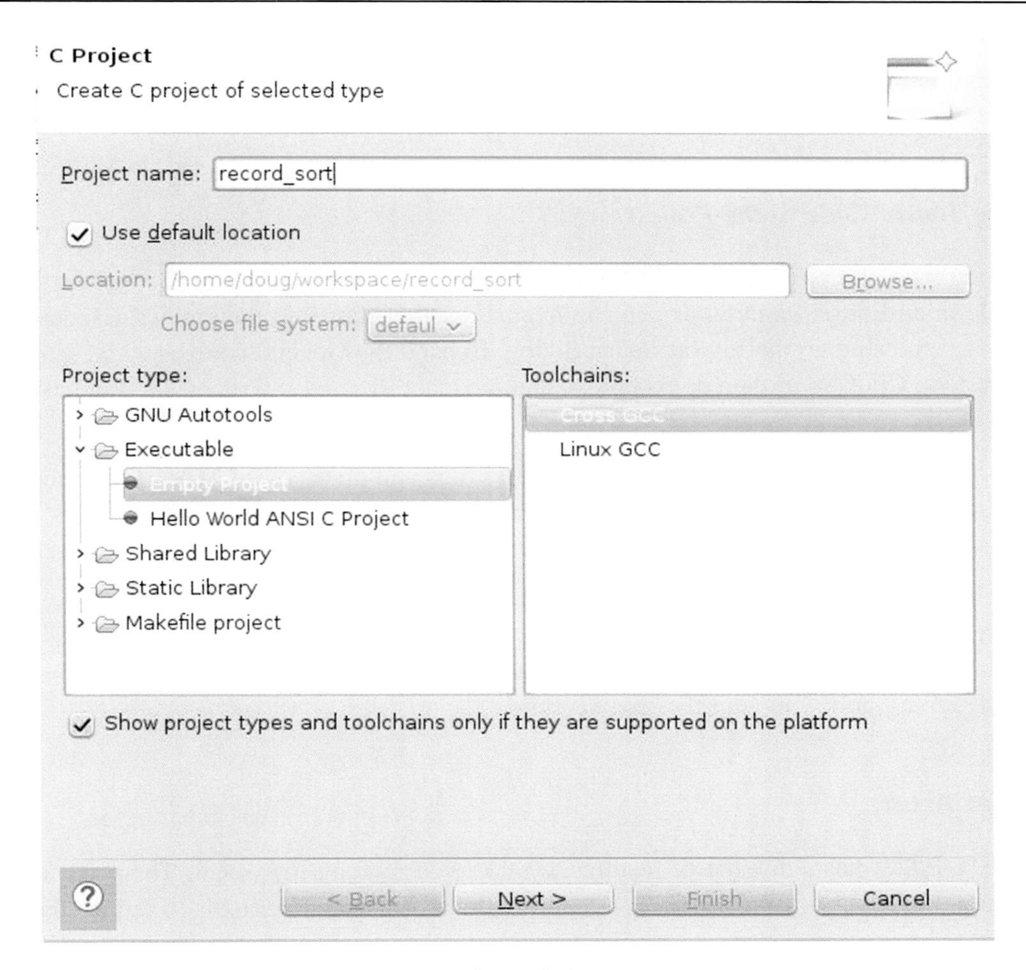


**Figure 5.4**
New Project wizard.

both of the Debug and Release configurations. Later you will have the choice of building either of these configurations. The primary difference between them is that the Debug configuration is built with the compiler's debug flag, "-g," turned on to provide symbol information to GDB. The Release configuration leaves out the debug flag (Fig. 5.4).

When you click `Finish` in the New Project wizard, you will find an entry in the Project Explorer view. The only item under the `record_sort` project is `Includes`, which is a list of paths to search for header files.

At this point, it would be useful to take a look at the directory `workspace/record_sort/`. It contains just two files, `.cproject` and `.project`, both of which are XML code describing the project. `.project` provides configuration information to the base Eclipse platform, while the more extensive `.cproject` file provides information to CDT. It is not necessary to understand the contents of these files, but it is useful to know they exist.

### Adding Source Code to the Project

There are basically two ways to add source code to a C project. You can of course create a new file in an Editor window, or you can *import* existing files into the project. Execute `File > Import...` to bring up the Import Select dialog. Expand the General category, and select File System. Click `Next`, then click `Browse`, and navigate to `samples/record_sort` and click `OK`.

This brings up the Import dialog. Select all three files and click `Finish`. Those files now show up in the Project Explorer view. Note that there is no `record_sort`.c file. That is because you are going to type it in yourself, to get some practice with the CDT editor.

Click the down arrow next to the New icon at the left end of the tool bar, and select `Source File` from the drop-down menu. Name it "record_sort.c." An Editor window opens up with a preliminary header comment. The contents of `record_sort.c` are given in Fig. 5.5, but do not start typing until you read the next section.

### Content Assist

The CDT Editor has a number of features to make your coding life easier. These fall under the general heading of *Content Assist*. The basic idea of Content Assist is to reduce the number of keystrokes you must type, by predicting what you are likely to type based on the current context, scope, and prefix. Content Assist is invoked by typing `Ctrl + Space`, and it is also auto-activated when you type ".," "->," or "::" following a `struct` or `class` name.

### Code Templates

*Code Templates* are an element of Content Assist that provide starting points for frequently used sections of code. When you enter a letter combination followed by `Ctrl + Space`, a list of code templates that begin with that letter combination is displayed. Select one and it is automatically inserted into the file at the current cursor location.

Try this: after entering the `#include` lines in Fig. 5.5, type "ma" `Ctrl + Space`. This brings up a template for the `main()` function. You can edit templates at `Window > Preferences > C/C++ > Editor > Templates`. Other aspects of Content Assist can also be customized under Preferences.

```
/*
 * author Doug Abbott
 *
 * Simple demonstration of building and running a project under
 * Eclipse.
 *
 * Usage:
 *        record_sort <filename> [1 | 2]
 *
 * Sorts records in <filename> either by name (1) or ID (2).
 * Default is name.  Outputs sorted file to stdout.
 */
#include <stdio.h>
#include <stdlib.h>

#include "record_sort.h"

int main (int argc, char **argv)
{
        int size, sort = 1;
        record_t *records;

        if (read_file (argv[1], &size, &records))
        {
                printf     ("Couldn't open file %s \ n", argv[1]);
                exit (1);
        }
        if (argc > 2)
                sort = atoi (argv[2]);

        switch (sort)
        {
                case 1: sort_name (size, records);
                        break;

                case 2: sort_ID (size, records);
                        break;

                default:
                        printf ("Invalid sort argument\n");
                        return_records (size, records);
                        exit (2);
        }
        write_sorted (size, records);
        return_records (size, records);
        return 0;
}
```

**Figure 5.5**
record_sort.c.

### *Automatic Closing*

As you type, note that whenever you type an opening quote ("), parenthesis ((), square ([) or angle (<) bracket, or brace ({), the Editor automatically adds the corresponding closing character and positions the cursor between the two. Type whatever is required in the enclosure, and hit `Enter`. This positions the cursor just beyond the closing character.

However, if you move the cursor out of the enclosed space, to copy and paste some text for example, the `Enter` key reverts to its normal behavior of starting a new line.

In the case of an opening brace, the closing brace is positioned according to the currently selected coding style, and the cursor is properly indented.

Finally, notice that as you type, appropriate entries appear in the Outline view identifying header files, functions, and, if we had any, global variables.

## The Program

Before moving on to building and running the project, let us take a closer look at what it actually does. `main()` itself is pretty straightforward. It is just a set of calls to functions declared in `sort_utils.c` that do the real work.

The function `read_file()` reads in a data file assumed to be organized as one record per line, where a record is a text name and a numeric ID. It allocates memory for an array of records, and a separate allocation for each name field.

There are two sort functions—one to sort on the name field, and the other to sort on the ID field. Both of these implement the shell sort algorithm, named after its inventor Donald Shell. Shell sort improves performance over the simpler insertion sort by comparing elements separated by a gap of several positions.

After the record array is sorted, `write_sorted()` writes it to `stdout`. This, of course, could be redirected to a file.

The final step is to return all of the allocated memory in the function `return_records()`.

The program does virtually no "sanity checking" and, if you are so inclined, you might want to build some in. There is also very little in the way of error checking.

## Building the Project

Once you have completed and saved the `record_sort.c` file, the next step is to build the project. All files that are created in, or imported into, a project automatically become a part of it, and are built and linked into the final executable.

In the Project Explorer view, select the top-level `record_sort` entry. Then execute `Project > Build Project` or right-click and select `Build Configurations > Build > All`. In the former case, the *Active Configuration* will be built. By default this is the Debug configuration. The Active Configuration can be changed by executing `Project > Build Configurations > Set Active`.

In the latter case, both the Debug and Release configurations will be built. In either case, one or two new entries will show up under `record_sort` in the Project Explorer view. These entries represent subdirectories called `Debug/` and `Release/` that hold, respectively, the object and executable files for the corresponding build configurations. Each also contains a makefile and some Eclipse-specific files.

Initially, the build will fail because some compile-time errors and warnings have been built into `sort_utils.c`. Open the Problems view, expand the Errors entry, and right-click on the item that says "record_t has no member named 'Id'." Select `Go To` to open `sort_utils.c`, if it is not already open, and highlight the line that has the error. The correction should be fairly obvious.

Eclipse CDT identifies build problems in several different ways. In the Project Explorer view, any project and source file that have problems are flagged with a red X icon for errors, or a yellow shield icon with a "!" to indicate a warning. When a source file with errors or warnings is opened in the Editor, the tab shows the same symbol. The Outline view then uses the same icons to show which functions in the file have either errors or warnings.

The Editor window uses the same icons to identify the line on which each error or warning occurs. You can scroll through a source file and click on a line displaying either a warning or error icon, and the Problems view will jump to the corresponding entry. If you roll the cursor over a line that is identified as an error or warning, the corresponding error message pops up.

Correct the problems and build the project again. Note, incidentally, that by default Eclipse does *not* automatically saved modified files before a build. You must manually save them. That behavior can be changed in Preferences. Save `sort_utils.c` and run the build again. This time it should succeed and you will see an executable file show up in the Debug tree in the Project Explorer view.

Before moving on to debugging, you might want to turn on line number display in the editor, since we will be referring to line numbers as we go along. In the editor window, right-click in the vertical bar on the left side, called the marker bar. One of the options is Show Line Numbers. Click that. This is a global parameter. It applies to any files currently open in the editor, and any that are subsequently opened.

## Debugging With CDT

CDT integrates very nicely with the Gnu Debugger, GDB. Like most Unix tools, GDB itself is command line oriented. Eclipse provides a very nice graphical wrapper around the native interface, so you don't have to remember all those commands.

Before we can debug, we have to create a *debug launch configuration.* Execute `Run > Debug Configurations...` Select `C/C++ Application` and click the `New` button in the upper left corner.

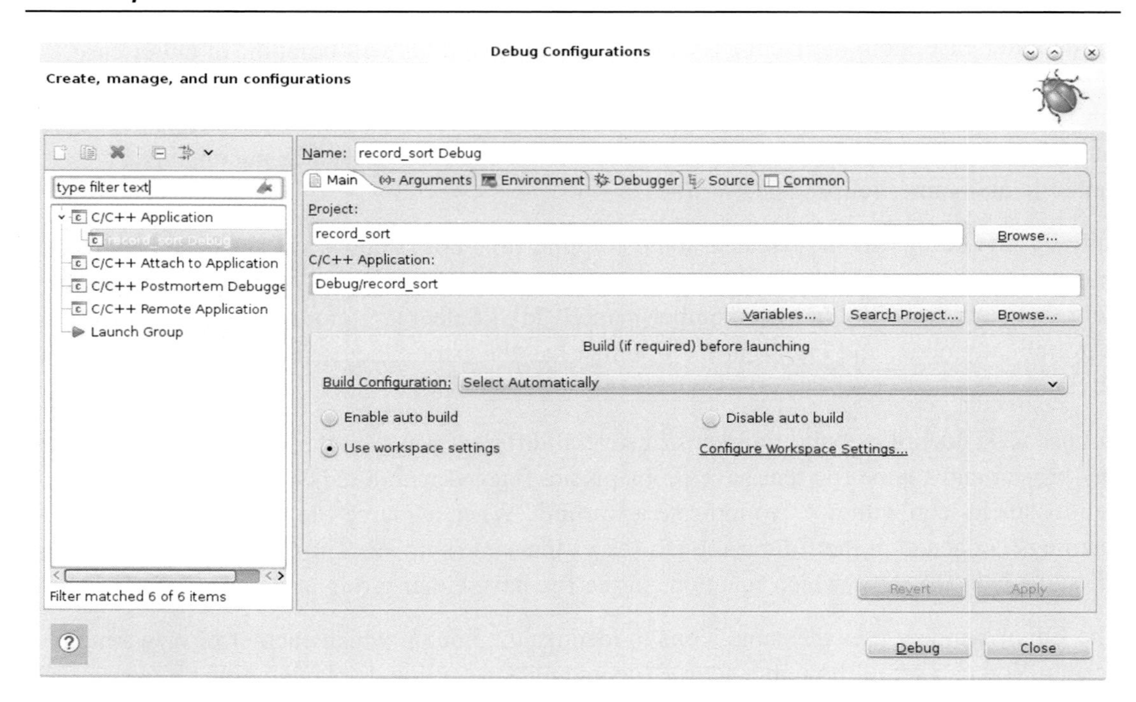


**Figure 5.6**
Debug Configuration dialog.

This brings up the Debug Configuration dialog shown in Fig. 5.6. Everything is already filled out correctly. We just need to make one addition. Select the Arguments tab and enter "datafile" into the Program arguments: window. `datafile` is a set of sample data records for the program to sort. It was imported into the project along with `sort_utils.c` and `record_sort.h`.

Click `Apply`, then `Debug`. You are asked if you want to open the Debug perspective. Yes, you do. You will probably want to check the box that says "Remember my decision," so you won't be bugged by this message again.

Fig. 5.7 shows the initial Debug perspective. Down at the bottom is a Console window where terminal I/O takes place. The Editor window shows `record_sort.c`, with the first line of `main()` highlighted. The Outline tab on the right lists all of the header files included by `record_sort.c`, and all of the global and external symbols declared in the file including variables, constants, and functions.

In the upper right is a set of tabs that provide a range of debug information including local variables, breakpoints, and registers. Initially, of course, all the local variables have random values since they are allocated on the stack. Just below the Debug tab in the upper left is a set of execution icons. Click on the Step Over icon. The highlight in the Editor window

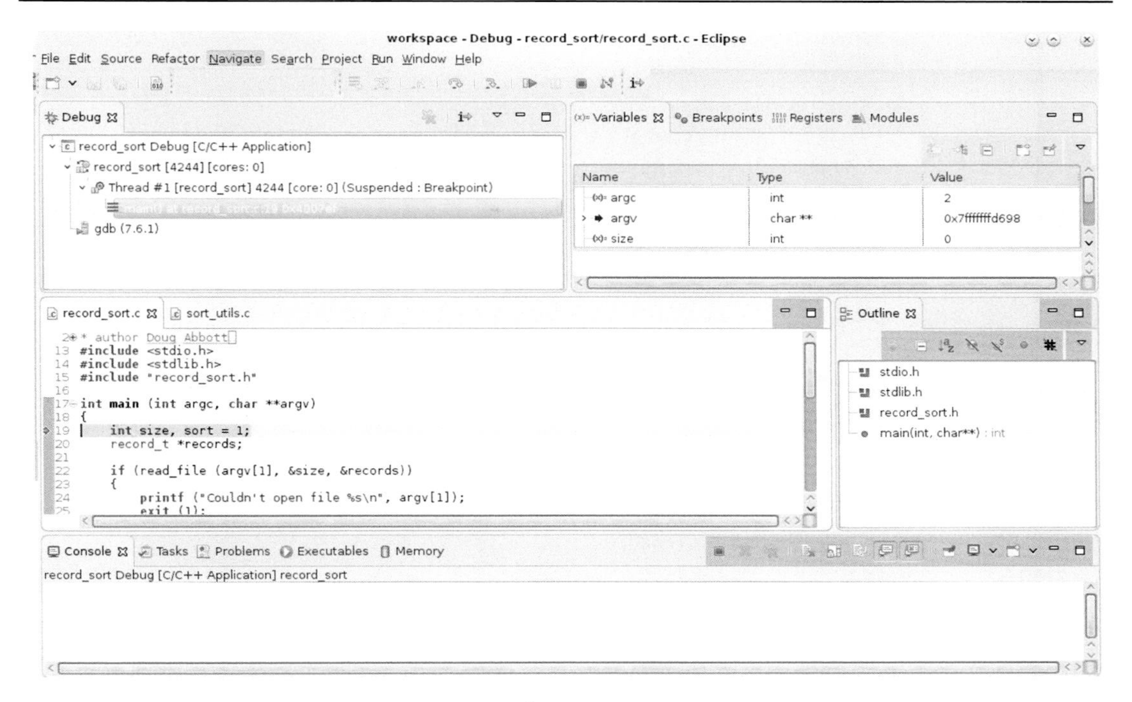


**Figure 5.7**
Debug perspective.

advances to the next line, and `sort` in the Variable view shows its initial value of 1. The value is highlighted in yellow, because it has changed since the last time the debugger was suspended. You can also see the current value of any variable by simply rolling the cursor over it. Give it a try.

For now, go ahead and let the program run by clicking the Resume icon in the main tool bar. Hmmm, we didn't get exactly the results we expected. `datafile` has 12 records, but only 1 record is output to the Console view. That is because a couple of run-time errors have been built into the program to offer some practice using the debugger.

### The Debug View

In the Debug view, right-click on the top level project entry and select `Relaunch` to start another debug run. The Debug view, shown in Fig. 5.8, displays the state of the program in a hierarchical form. At the top of the hierarchy is a *launch instance*, that is an instance of a launch configuration identified by its name. Below that is the debugger instance, identified by the name of the debugger, in this case gdb. Beneath the debugger are the program threads under the debugger's control. For `record_sort` there is just one thread. Later on we will see how gdb/Eclipse handles multithreaded programs.

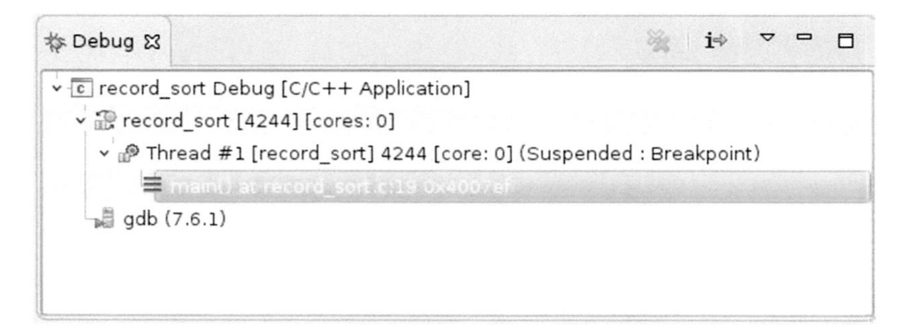


**Figure 5.8**
Debug view.

Finally, at the lowest level are the thread's stack frames, identified by function name, source code line, and program counter. Currently, there is only one stack frame for `main()` stopped at `record_sort.c`, line 22.

The main tool bar has several buttons for managing program execution. Roll your cursor over them to see the labels.

Click `Step Over` once, and then click `Step Into` to get into the `read_file()` function. Note that a second stack frame appears in the Debug view, and `sort_utils.c` is opened in the Editor. At this point it would be worth taking a closer look at the four tabbed views in the upper right of the workbench.

### *Variables View*

When a stack frame is selected in the Debug view, the Variables view displays all the local variables in that frame. Right now you should be in the `read_file()` function. The two variables visible are both pointers. Clicking the white arrow to the left of the name dereferences the pointer, and displays the corresponding value. For string variables, the full string is displayed in the lower window.

Select the `main()` stack frame in the Debug view, and note that the Variables view changes to show the local variables in `main()`. If anything other than a stack frame is selected, the Variables view goes blank. Remember, you can also view the value of a variable simply by rolling the cursor over it.

### *Breakpoints View*

To debug the problems in record_sort, you will probably want to set one or more breakpoints in the program and watch what happens. Select the Breakpoints view, which is currently empty because we haven't set any breakpoints.

Let us set a breakpoint at line 34 in `sort_utils.c`. That is the beginning of an if statement in `read_file()`. Right-click in the marker bar at line 34, and select `Toggle Breakpoint`. A green circle appears to indicate that an enabled breakpoint is set at this location. The check mark indicates that the breakpoint was successfully installed.

A new entry appears in the Breakpoints view, with the full path to the breakpoint. The check box shows that the breakpoint is enabled. Click on the check box to disable the breakpoint, and note that the circle in the marker bar changes to white. Disabled breakpoints are ignored by the debugger. Click the check box again to re-enable it.

Click `Resume` in the Debug view tool bar, and the program proceeds to the breakpoint. The Thread [0] entry in the Debug view indicates that the thread is suspended, because it hit a breakpoint. Click `Step Over` to load `temp` with the first record. Select the Variables view and click the white arrow next to `temp`. Now you can see the current values of the fields in `temp`. Variables whose value has changed since the last time the thread was suspended are highlighted in yellow.

Breakpoints have some interesting properties that we will explore later on.

## *Memory View*

There is one more debug-oriented view that shows up by default in the bottom tabbed window of the Debug perspective. The Memory view lets you monitor and modify process memory. Memory is organized as a list of *memory monitors*, where each monitor represents a section of memory specified by a base address. Each memory monitor can be displayed in one or more of four predefined formats known as *memory renderings*. The predefined renderings are hexadecimal (default), ASCII, signed integer, and unsigned integer.

Fig. 5.9 shows a memory monitor of the area allocated for `temp` just after the first `fscanf()` call in `read_file()`. The Memory view is split into two panes, one that lists the currently active monitors and another that displays the renderings for the selected

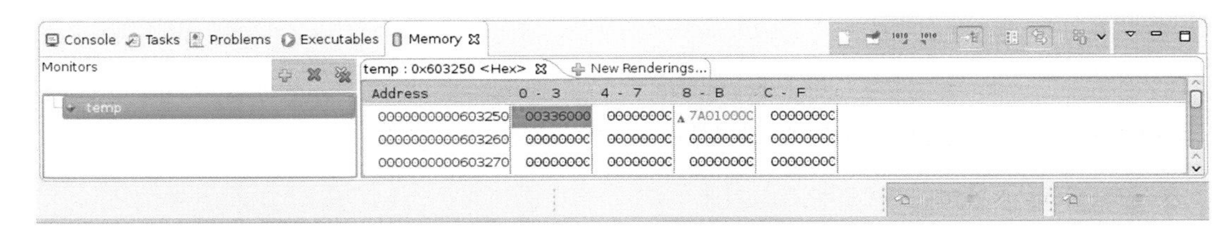


**Figure 5.9**
Memory view.

monitor. A monitor may have multiple renderings enabled, and these will be tabbed in the renderings pane.

The first four bytes hold a pointer to another area allocated for the name. Remember that the x86 is a "little endian" machine. Consequently, when memory is displayed as shown here, most entries appear to be "backwards."

Each of the panes has a small, fairly intuitive set of buttons for managing the pane. These allow you to either add or remove monitors or renderings and, in the case of monitors, remove all of them.

## Finish Debugging

With this background on the principal debugging features of Eclipse, you should be able to find the two runtime errors that have been inserted in `sort_utils.c`. Good luck.

## Additional Plug-ins

There are some additional plug-ins that we will need later for debugging on the target system.

Go to the `Help` menu and select `Install New Software`. That brings up the dialog shown in Fig. 5.10. In the `Work with` dropdown box, select the entry corresponding to your version of Eclipse. In my case it's `Neon -` http://download.eclipse.org/releases/neon. Scroll down to `Mobile and Device Development` and expand that entry. Check the boxes for:

- `C/C++ Remote Launch (Requires RSE Remote System Explorer)`
- `Remote System Explorer End-user Runtime`
- `Remote System Explorer User Actions`
- `Target Management Terminal (Core SDK)`
- `TCF Remote System Explorer add-in`
- `TCF Target Explorer`

If these selections do not appear in the list, it means the items are already installed. Expand `Programming languages` and check the boxes for:

- `C/C++ Autotools Support`
- `C/C++ Development Tools`

Click `Next` to see a review of the items to be installed. Click `Next` again, to review and accept the terms of the license. Click `Finish` and the installation commences. When it finishes, Eclipse will tell you that it needs to restart.

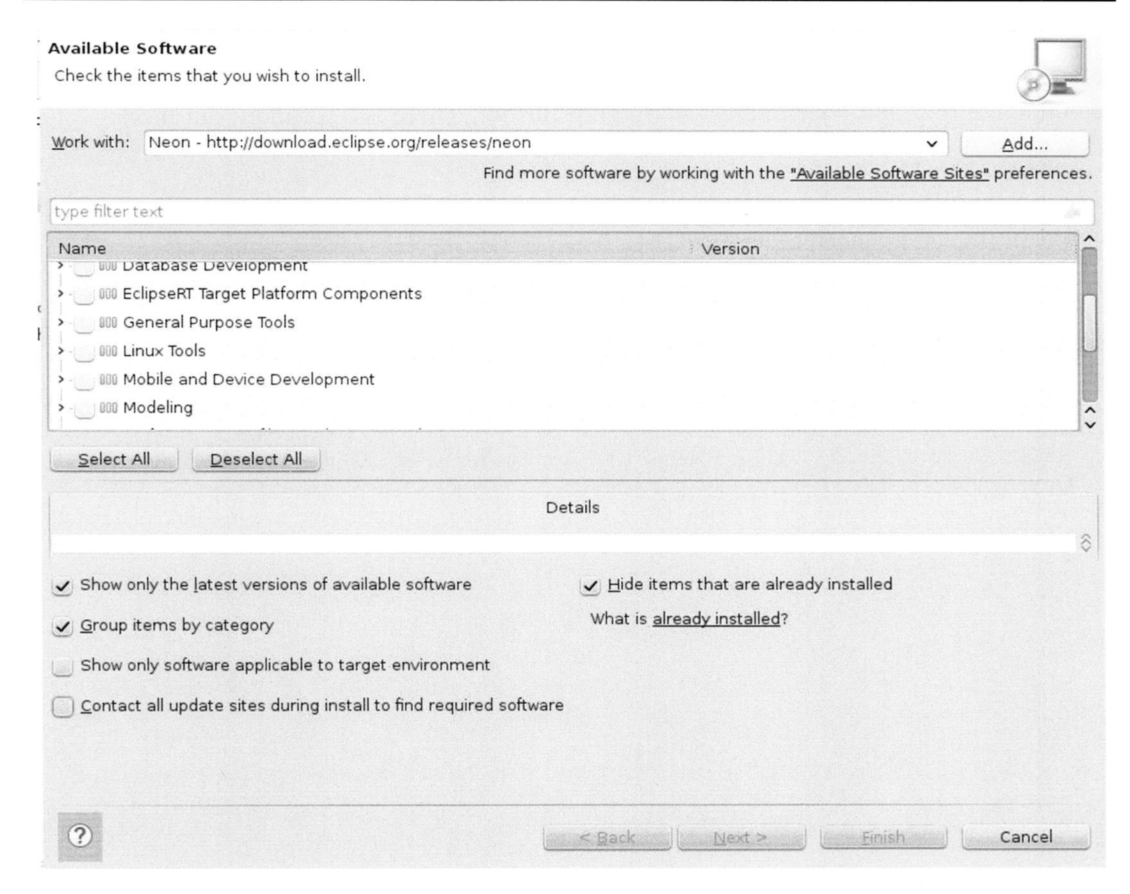


**Figure 5.10**
Eclipse Software Installation Dialog.

## *Summary*

In my humble opinion, Eclipse is the most professionally executed Open Source project I've seen. It is well thought out, and meticulously implemented. While it has tremendous power and flexibility, the most common operations are relatively simple and intuitive. The documentation, while still reference in nature and not tutorial, is nevertheless readable, and for the most part accurate. Sadly, there are all too many Open Source projects for which that cannot be said.

The plug-in mechanism makes Eclipse infinitely extensible. This opens up opportunities for both Open Source and commercial extensions to address a wide range of applications. The Resources section below lists a repository of Eclipse plug-ins.

Like many of the topics in this book, we've really just skimmed the surface, focusing on those areas that are of specific interest to embedded developers. But, hopefully, this has been enough to pique your interest in digging further. There is a lot more out there. Go for it!

We are now fully prepared to do some real programming for the target board. That is the subject of the next chapter.

## Resources

www.eclipse.org – The official website of the Eclipse Foundation. There is a lot here and it is worth taking the time to look through it. I particularly recommend the C/C++ Developer Toolkit page.

http://marketplace.eclipse.org/ – Eclipse Plugins. This page lists well over 1000 plug-ins, both commercial and Open Source.

Abbott, Doug, *Embedded Linux Development Using Eclipse*, Elsevier, 2008. OK, this one is a little old, but it does go into much more detail about using Eclipse for embedded development.

CHAPTER 6

# The hardware

## Chapter Outline



*Beware of programmers carrying soldering irons.*

***Anonymous***

Embedded computing covers an extremely wide range of hardware, from the (probably) 8-bit processor that powers your automatic coffee maker, to the (perhaps multicore) 32-bit or 64-bit processor in your cell phone or tablet, up to the massively parallel processing environment that runs the telecommunications network. In this chapter, we will get to work on the embedded target board that we will use in the next few chapters for applications programming.

## The ARM Architecture

ARM is a 32-bit and 64-bit reduced instruction set computer (RISC) architecture developed by ARM Holdings, a British company originally known as Advanced RISC Machines. It is said to be the most widely deployed 32-bit architecture in terms of numbers produced.

Linux for Embedded and Real-time Applications.
DOI: http://dx.doi.org/10.1016/B978-0-12-811277-9.00006-7

According to Wikipedia, over 50,000,000,000 ARM processors had been produced as of 2014.

The basic idea behind RISC architecture is that a simpler instruction set requires fewer transistors, which in turn consume less power and take up less die space. The relative simplicity of the RISC architecture makes ARM processors suitable for low power applications. As a result, they have come to dominate the mobile and consumer electronics market. About 98% of all mobile phones use at least one ARM processor[1].

ARM Holdings itself does not make chips. Rather, it licenses the architecture to a number of semiconductor manufacturers including Applied Micro, Atmel, Freescale, Marvell (formerly Intel), Nvidia, ST-Ericsson, Samsung, and Texas Instruments, among others.

## Open Source Hardware

Open source software has been around for at least 25 years. More recently, some groups have started developing open source hardware, generally small, single board computers based on ARM processor chips.

### BeagleBoard

The BeagleBoard is based on the TI (Texas Instruments) OMAP3530 system-on-a chip (SOC) that combines an ARM Cortex-A8 CPU with a TI TMS320C64x digital signal processor. It was initially developed by TI as an educational tool, and the design was released under the Creative Commons license. The original BeagleBoard was release in July of 2008 (Fig. 6.1).

#### Specifications (Rev. C4)

- Processor: TI OMAP3530 at 720 MHz
  - ARM Cortex-A8
  - TMS320C64x + DSP
- 256 MB RAM
- 256 MB NAND flash
- Peripheral connections:
  - DVI-D using HDMI connector, max. resolution 1280 × 1024
  - S-video
  - USB 2.0 HS OTG
  - RS-232

[1] c|net, "ARMed for the Living Room," http://news.cnet.com/ARMed-for-the-living-room/2100-1006_3-6056729.html.

**Figure 6.1**
BeagleBoard.

  - JTAG
  - SD/MMC slot
  - Stereo audio in and out
  - Power: 5 VDC, barrel connector
- Size: 75 mm (~3 inches) × 75 mm
- Suggested distributor price: $125

Requiring only 2 W at 5 V, the board can also be powered through its USB port.

There have been several versions of the BeagleBoard released over the years. These include:

*BeagleBoard-xM*

- Processor: TI DM3730 at 1 GHz (Fig. 6-2)
  - ARM Cortex-A8
  - TMS320C64x + DSP

**Figure 6.2**
BeagleBoard-xM.

- 512 MB RAM
- 4 GB Micro SD card with Angstrom Linux distribution
- Peripheral connections:
    - DVI-D using HDMI connector, max. resolution 1280 × 1024
    - S-video
    - USB 2.0 HS OTG (mini AB)
    - 4 USB ports
    - Ethernet
    - RS-232
    - JTAG

  - SD/MMC slot
  - Stereo audio in and out
  - Camera port
  - Expansion port
  - Power: 5 VDC, barrel connector
- Size: 82.5 mm (~3.25 inches) × 82.5 mm
- Suggested distributor price: $149

*BeagleBone*

A "bare bones" version (Fig. 6-3). The BeagleBone has a pair of 46-pin expansion connectors into which expansion boards called "capes" may be inserted. Up to four capes can be stacked at once. A large number of capes are available. See the Resources section for a link to a page listing them.

- Processor: TI AM335x ARM Cortex-A8 at 720 MHz
- 256 MB RAM
- 4 GB Micro SD card with Angstrom Linux distribution
- Peripheral connections:
  - USB Host and Client
  - Ethernet
  - RS-232 and JTAG on 10-pin header
  - Micro SD slot
  - Power: 5 VDC, barrel connector
- Size: 86 mm (~3.4 inches) × 53 mm (~2.1 inches)
- Suggested distributor price: $89

*BeagleBone Black*

Successor to the BeagleBone. The Black offers more features at a lower price. The hardware-related exercises in the book are designed for either the BeagleBone or the BeagleBone Black (Fig. 6.4).

- Processor: TI AM335x ARM Cortex-A8 at 1 GHz
- 512 MB RAM
- 2 (Rev. B) or 4 (Rev. C) GB eMMC flash. Rev. B ships with Angstrom, Rev. C ships with Debian
- Peripheral connections:
  - USB Host and Client
  - Ethernet
  - Micro SD slot
  - Micro HDMI
  - TTL serial port on 6-pin header

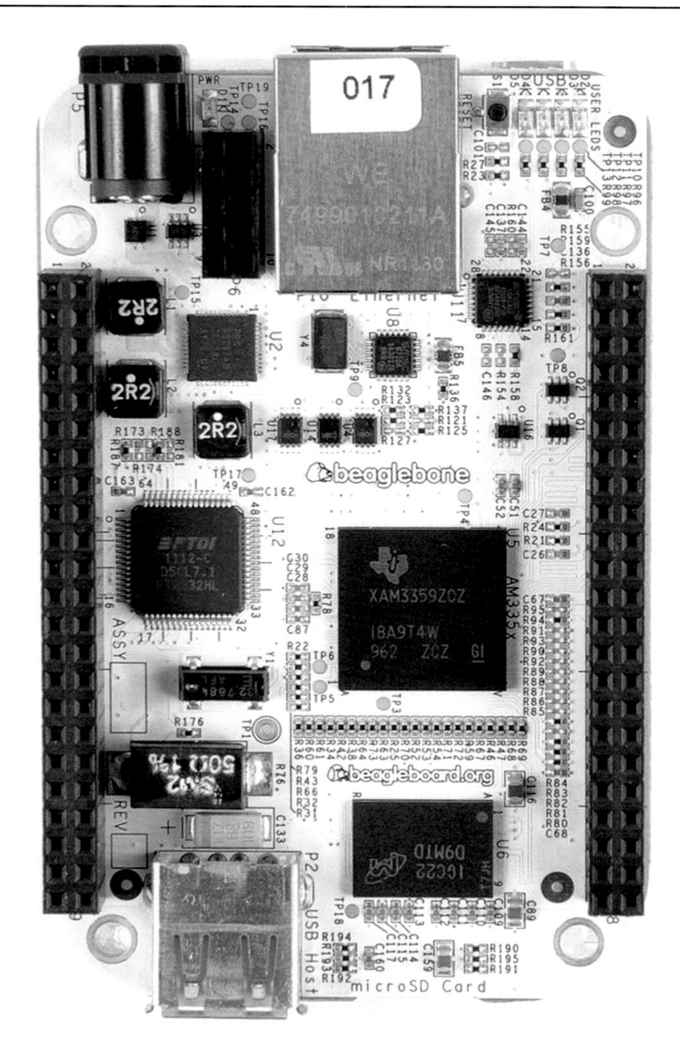


**Figure 6.3**
BeagleBone.

  - JTAG on 20-pin header (unpopulated)
  - Power: 5 VDC, USB, barrel connector, or battery
- Size: 86 mm (~3.4 inches) × 53 mm (~2.1 inches)
- Suggested distributor price: $60

We will use the BeagleBone Black to explore embedded Linux for much of the rest of the book. Some of the exercises require a potentiometer connected to an analog input. The easiest way to do that is with a prototyping cape, such as the one offered by Sparkfun (see Resources for web address). You will also need to order two 46-pin headers.

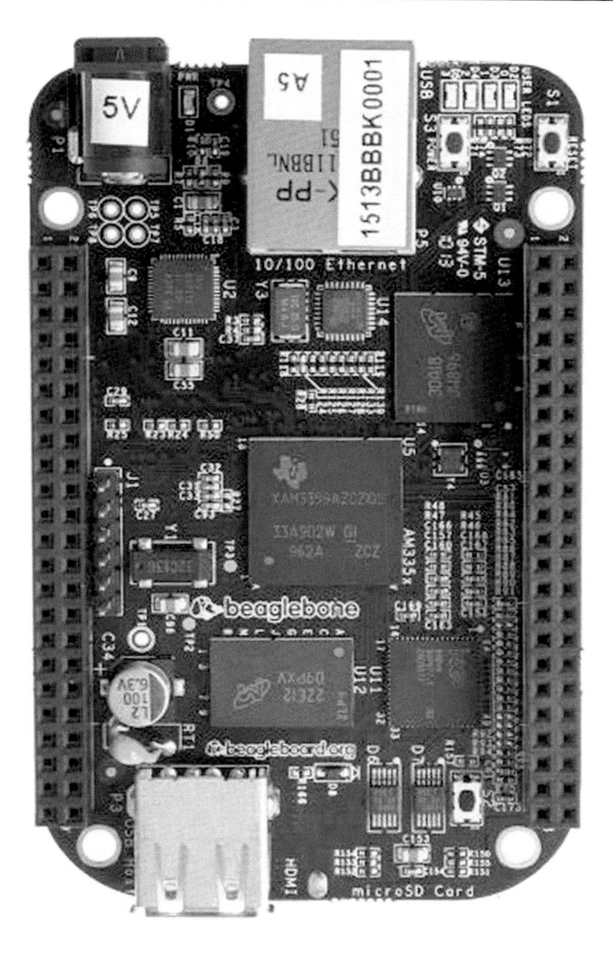


**Figure 6.4**
BeagleBone Black.

The other accessory you may want is a USB to serial TTL cable. While the BeagleBone Black does emulate a serial port on its USB client, that does not access the u-boot bootloader, which we will need to make some modifications to. Again, there are multiple sources for this cable, among them Sparkfun.

### *Gumstix*

The Gumstix series of SBCs are produced by a company of the same name. The name comes from the size of the boards, which is about the size of a stick of chewing gum. There are currently three product lines:

- Overo series based on the TI OMAP ARM Cortex A8 processors
    - 512 MB RAM

  - 1 GB NAND flash
  - 11 models priced from $109 to $179
- DuoVero series based on TI OMAP4430 ARM Cortex A9 dual core processors
  - 1 GB RAM
  - No built in flash
  - Three models priced from $169 to $199
- Verdex Pro series based on Marvell PXA270 Xscale processors
  - 128 MB RAM
  - 32 MB NOR flash
  - Three models priced from $129 to $169

Owing to its small size, the computer board itself does not have any standard I/O connectors. Connection to the board is through two small 70-pin AVX connectors that mate with an expansion board that provides the usual connections (Fig. 6.5).

**Figure 6.5**
Gumstix Overo.

*Additional specifications*

- Bluetooth and 802.11 G wireless available on some models
- On-board MicroSD slot
- Size: 58 mm (~2.3 inches) × 17 mm (~0.67 inches)

Several expansion boards are available with a wide range of connection options including Ethernet, USB, stereo audio in and out, camera interface, and DVI video, among others. The expansion boards are priced from $27.50 to $229.

## *Raspberry Pi*

In March of 2012 the big buzz in the world of ARM-based SBCs was the Raspberry Pi, a $35 board that runs standard Linux distributions such as Fedora and Debian. Developed in the UK by the Raspberry Pi Foundation, the objective is to provide a low-cost computer that will stimulate the teaching of computer science in schools.

The device went on sale February 29, 2012, and almost immediately the initial production run of 10,000 units sold out.

Like all of the open source boards we have looked at, the RPi has gone through a substantial evolution. The latest revision, called the Raspberry Pi 3 Model B, was released in February of 2016 (Fig. 6.6).

**Figure 6.6**
Raspberry Pi 2.

*Specifications*

- Processor: Quad core ARM Cortex-A53, Broadcomm BCM2837 at 1.2 GHz
- 1 GB of RAM
- No flash, boots from SD card
- Peripheral connections
    - Ethernet
    - WiFi
    - Bluetooth
    - USB 2.0

  - HDMI
  - RCA video
  - Stereo audio out
  - Header footprint for camera connection
- Powered through microUSB socket
- Size: 85.6 mm (~3.4 inches) × 56.5 mm (~2.2 inches)

## Setting up the BeagleBone Black

Setting up the BeagleBone Black is pretty straightforward. The Quick Start guide included with the board is really about all you need to get started. So, start by connecting the board to your workstation using the supplied USB cable. Initially, the four blue user LEDs will come on solid. After about 10 seconds, they will assume their default assigned function:

- USR0 – Heartbeat
- USR1 – microSD access
- USR2 – CPU activity. That is, the kernel is not idle
- USR3 – eMMC activity

Fig. 6.7 shows the layout of the board with specific reference to connectors, switches, and indicators. The only cable supplied with the board is a USB cable. The simplest setup is to

**Figure 6.7**
BeagleBone Black layout.

simply connect the board to your workstation through the USB cable, which supplies up to 500 mA of power for the board.

The board's USB slave port is a multifunction device that appears as:

- Mass storage device that appears at `boot` under the normal path for removable devices (`/run/media/<your_user>` in CentoOS 7)
- Serial port. Device name is `/dev/ttyACM0`
- Network device with IP address 192.168.7.2.

Mount the board's mass storage device, and scan through the `START.htm` page in Firefox. Among other things, it will instruct you to execute the script file `mkudevrule.sh`. This is necessary in order to utilize the board's serial and network ports. It would probably be a good idea to reboot your workstation after this, to make sure that the changes have been implemented.

Now fire up `minicom`. You should see a screen like Fig. 6.8. The board gives you a normal user password, or you can log in as root with no password. Try a few shell commands just to prove it really is a Linux shell. Execute `uname -a` to verify that you really are talking to a BeagleBoard.

```
Debian GNU/Linux 7 beaglebone ttyGS0

default username:password is [debian:temppwd]

Support/FAQ: http://elinux.org/Beagleboard:BeagleBoneBlack_Debian

The IP Address for usb0 is: 192.168.7.2
beaglebone login: █
```

**Figure 6.8**
BeagleBoard serial port.

Finally, in another shell window, execute `ssh -l debian@192.168.7.2`. You will be asked for the same password as the minicom shell. Or, again, you can log in as root with no password.

So, you have two options for getting shell access to the target board: the serial port, or SSH over the network. Both of these operate over the same USB interface.

You will also want to connect your BBB to your network through the Ethernet port, so you can Dynamic Host Configuration Protocol (DHCP) an IP address.

## Flash Memory and File Systems

### BeagleBone Black Memory Configuration

The BeagleBone Black has several forms of memory, both volatile and nonvolatile, as shown in Fig. 6.9. There are two forms of nonvolatile storage: a 4 KB EEPROM (Electrically Erasable Programmable Read-Only Memory) that stores ID data about the board itself, and a 4 GB eMMC (embedded Multi-Media Card) NAND flash that uses a byte-wide port for address and data. In addition to the on-board NAND flash, there is a microSD card slot for additional storage. Finally, there is 512 MB of DDR3 RAM for working memory.

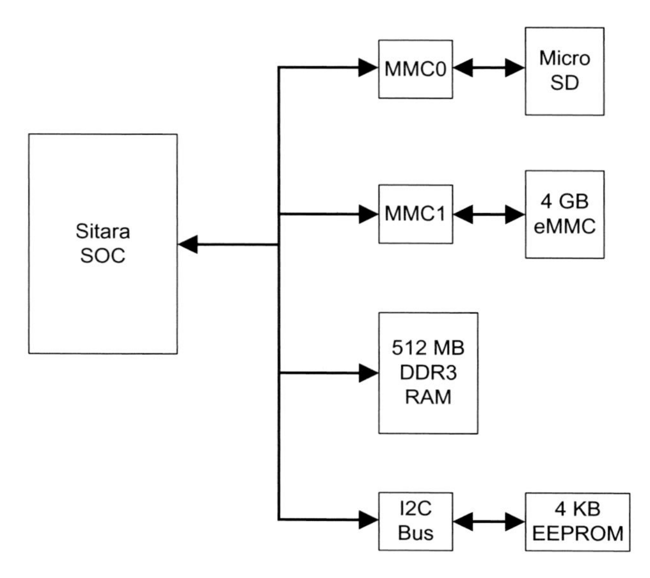


**Figure 6.9**
BeagleBone Black memory configuration.

The eMMC is divided into two partitions: a 32 MB VFAT formatted boot partition and the remainder is formatted as an ext4 root file system. Once the board is booted, the boot partition is visible to your workstation as a USB drive. It is also mounted in the BBB's file system as `/boot/uboot`.

## Preparing the Board

### Caveat

In the remainder of the book, I'll be describing Debian version 7.4 released in 2014. There are two reasons for this:

1. It was installed on the board I bought back then;
2. It works.

I tried the latest release, Debian 8.7, but ran into problems that I couldn't resolve and couldn't find help with on-line. Specifically, I couldn't mount the root file system over Network File System (NFS), although all of the documentation says it should work fine. Clearly, after you finish the book, you're free to install whatever version strikes your fancy. But for now, we are going to stick with version 7.4.

We are going to make some changes to the way the BBB boots up. Specifically, instead of mounting the root file system from the eMMC, we are going to mount it over NFS from the workstation. This requires modifying the boot process.

### *The Boot Process*

The BBB uses a very capable open source boot loader called u-boot, which we will get into in more detail in a later chapter. U-boot makes extensive use of environment variables to control the boot process.

Hook up your serial cable per the instructions that came with it. Change the `minicom` Serial Device from `ttyACM0` to `ttyUSB0`. Reset the board, but this time quickly hit a key. I usually hit the space bar. This brings up the u-boot prompt, along with some basic board information as shown in Fig. 6.10.

Enter the command `printenv` to see the environment variables. There are a lot of them. Environment variables can be set with the setenv command:

```
setenv variable value
```

or by entering:

```
variable=value
```

```
U-Boot 2014.04-00014-g5a2dca1 (Apr 14 2014 - 17:00:58)

I2C:   ready
DRAM:  512 MiB
NAND:  0 MiB
MMC:   OMAP SD/MMC: 0, OMAP SD/MMC: 1
*** Warning - readenv() failed, using default environment

Net:   <ethaddr> not set. Validating first E-fuse MAC
cpsw, usb_ether
Hit any key to stop autoboot:  0
U-Boot# █
```

**Figure 6.10**
u-boot prompt.

The value of an environment variable is referenced by enclosing it in `${}`. An environment variable can contain one or more u-boot commands separated by ';'. Such a variable can be "executed" with the `run` command. A variable can also contain the `run` command, allowing it to execute other variables.

U-boot has a very extensive command set, much of which is detailed in Appendix A. For now, type help to see a complete list.

Note the message:

```
*** Warning - readenv() failed, using default environment
```

U-boot tries to read its environment from a NAND flash that is not installed on the BBB. That fails, and so it uses a default environment built into u-boot itself. One consequence of this is that we cannot permanently modify the default environment without modifying and rebuilding u-boot itself. We can, of course, modify variables each time we boot.

But note the variable `bootenv=uEnv.txt`. When u-boot has finished setting up the default environment, it reads the file specified by `bootenv` for additional instructions. This file resides in the eMMC's boot partition that is visible in your workstation.

Issue the `boot` command to let Linux boot up. Mount the boot partition, and open `uEnv.txt` with an editor. The interesting piece is down at the end, the variable `uenvcmd`. This is the last variable u-boot executes, and so it should boot up the kernel in some fashion.

In the default case, `uenvcmd` loads three files from the eMMC: the kernel, the initial RAM disk image (initrd), and something called a *flattened device tree* (*fdt*). `mmcargs` sets the `bootargs` variable, which is passed to the kernel as it boots. The `bootz` command transfers control to the kernel, passing it the addresses of the initrd and fdt. So, to modify how the root file system gets mounted, we modify `uEnv.txt`, or perhaps better, supply a different one.

In your samples/ directory, open the file `uEnvnet.txt` and find `uenvcmd`. About the only thing that is different is it runs `netargs` instead of `mmcargs`. `netargs` sets up `bootargs` to specify the file system type as nfs, and provides the location of the server and path to the file system directory. Up around the middle of the file is a set of variables that define the address of the server and the path to the root file system, which you will probably want to change.

Historically, I've always defined a fixed IP address for the target board. For some reason that does not work with the BBB, and so you see at the tail end of `netargs ip=dhcp`.

In principle, you should be able to write directly to the boot partition from your workstation, but I found that to be less than reliable. What I did was to create a directory on the BBB under `/home`, and then mount my workstation's home directory there over NFS.

But, before you can do that, you have to install a package called `nfs-common`. Your BBB must be connected to the Internet. Then execute:

```
sudo apt-get update
sudo apt-get install nfs-common
cd /home
sudo mkdir doug
sudo mount –t nfs –o nolock 192.168.1.2:/home/doug doug
```

Of course, if you are logged in as root, you can drop the sudo. You might want to put the `mount` command into a shell script, as you may be executing it a number of times. Now you can copy `uEnvnet.txt` from the workstation to `/boot/uboot` on the BBB:

```
cp doug/samples/uEnvnet.txt /boot/uboot
sync
```

Always execute `sync` after changing the `/boot/uboot` directory, or your changes may not get to the flash before you reboot. I suggest you do not flat out replace the original `uEnv.txt` just yet and in fact, you might want to keep a copy of it.

### Differences Between Version 7.4 and Version 8.7

Now that we understand the boot process, I can say a little about the differences between Debian 7.4 and the latest release, 8.7. The principal difference I see between the two is in how the boot code is structured. As described above, version 7.4 splits the eMMC into two partitions, one for booting and one for the root file system. In version 8.7, there is only the ext4 root partition that includes the kernel image, the initrd, and device table files. The u-boot image is not visible. It is clearly there, but it is hidden.

The default u-boot environment has expanded substantially. In 7.4 it is 4.6 KB, and in 8.7 it is 20.5 KB. The intention is to make it easier to write `uEnv.txt` files, which is fine, but it also makes the default environment itself much harder to understand.

### The Root File System Image

Even though we will be mounting the root file system over NFS, you will need to load an *image* file to the eMMC in order to create a boot partition. But first you need to load the image file to a microSD card. A 4 GB card is sufficient. You should use a USB card reader, as there have been reports that the built-in readers on some laptops are not reliable under Linux. The card should show up as `/dev/sdb`. If not, you can execute:

```
dmesg | tail
```

just after inserting the card, and it should show what device was just recognized.

Browse to debian.beagleboard.org/images and download `BBB-eMMC-flasher-debian-7.4-2014-04-14-2gb.img.xz`. Uncompress the file with the `xz` command thus:

```
xz –d BBB-eMMC-flasher-debian-7.4-2014-03-26-2gb.img.xz
```

This yields one file, `BBB-eMMC-flasher-debian-7.4-2014-03-26-2gb.img`. Write this file to the microSD card with the dd command:

```
dd if=BBB-eMMC-flasher-debian-7.4-2014-03-26-2gb.img of=/dev/sdb
```

This will take a fairly long time. When it is finished, there will be two partitions on `/dev/sdb`; a bootable FAT partition named BEAGLEBONE, and an ext4 partition named eMMC-Flasher. This image is specifically for flashing the eMMC.

Insert the microSD into your unpowered BBB. The BBB documentation says that to boot from the microSD card, you must hold down the boot pushbutton while applying power. In my case, the mere presence of a microSD card causes the board to boot from the card. Power it up. The kernel will boot up as normal, and present a login prompt. But the pattern of the LEDs indicates that something else is going on. That something else is flashing the eMMC. When the four LEDs come on steady, the flashing is finished. Power down the board, remove the microSD card and repower the board. It boots Debian 7.4 from the eMMC.

Finally, you need to get the target root file system on to your workstation under your home directory. You do not want to copy it from the microSD card, because that image is specifically intended for flashing the eMMC. What you can do is copy from the BBB eMMC over the network to your workstation. Create a subdirectory under your home directory to match what you specified for `rootpath` in `uEnvnet.txt`. Now copy the entire target file system into that directory. I suggest copying each top level directory separately, rather than everything in one big `cp –r` command. Note, incidentally, that there are several top level directories that are empty. You can just execute a `mkdir` command for these. A few directories require some special attention:

- `/boot` – Do not copy the `/uboot` directory. Just create an empty directory.
- `/dev` – You only want the part of `/dev` that is on the microSD card. The remainder of `/dev` is populated by a program called `udev`.
- `/home` – Do not copy your home directory that you mounted over NFS.

You will need to edit `/etc/fstab` as root user. There will be a line that looks something like this:

```
UUID=3bb11f04-4b37-49fd-a118-f1a7c1c4cfd7 / ext4 noatime,errors=remount-ro 0 1
```

Your UUID will probably be different. This line is trying to mount the root file system from a disk drive, which not surprisingly interferes with mounting over NFS. Just comment the line out and save the file.

### Boot to the NFS File System

Reset the board into the u-boot prompt if you are not already there. Execute the following commands:

```
setenv bootenv uEnvnet.txt
boot
```

The reason for not overwriting `uEnv.txt` right now is that there might be something wrong with your `uEnvnet.txt`, in which case you will need to boot back into the eMMC file system. When you are confident that `uEnvnet.txt` is working correctly, you can rename it `uEnv.txt` and boot directly into the NFS file system.

Interestingly, the boot up seems to be somewhat more verbose with the NFS file system than with the eMMC file system.

As a final step, move the `samples/` directory into your target file system under `/home`.

### What Can Go Wrong?

It is not unusual to encounter difficulties when bringing up an embedded target board such as the BeagleBone Black, even though the designers have gone to great lengths to make the process as painless as possible. The most common problems fall into two broad categories. The first is the serial port. Make sure the baud rate is set correctly. This is generally not a problem, because 115 kbaud seems to be the default for minicom. A more common problem is not turning off hardware flow control.

Common networking problems include having Security Enhanced Linux enabled, and/or the firewall turned on. This will prevent the target from NFS mounting its root file system. As a starting point, disable SELinux, and turn off the firewall. Later on you can configure either of these features to allow NFS mounting, but still provide some level of protection.

Make sure the IP address is correct. Actually, this should not be a problem, because you will be getting an IP address through DHCP.

With the target board fully configured and connected to the workstation, we can turn our attention to the task of writing applications. That's the subject of the next chapter.

## Resources

beagleboard.org – web site for the BeagleBoard series of boards
debian.beagleboard.org/images – A collection of root file system images for BeagleBoards.

www.dropbox.com/s/d7w56b980pakzzq/BBB-eMMC-flasher-debian-7.4-2014-04-14-2gb.img.xz?dl = 0 – In case the version 7.4 flasher image disappears from the debian web page, it is also here.

groups.google.com/forum/#!forum/beagleboard – Google group focused on the BeagleBoards

elinux.org/Beagleboard:BeagleBoneBlack_Debian – A page on the Embedded Linux wiki

*Sites for Alternate Boards*

Gumstix: gumstix.org, gumstix.com

Raspberry Pi: raspberrypi.org

PART II

# Application programming in a cross-development environment

CHAPTER 7

# Accessing hardware

## Chapter Outline



> *The only people who have anything to fear from free software are those whose products are worth even less.*
>
> ***David Emery***

## Review

This is a good time to stop, take a deep breath, and see where we have been before we move on to tackle application development.

Chapter 1, The embedded and real-time space, gave us an overview of the embedded and real-time space, and how Linux fits into that space. Chapter 2, Installing Linux went through the process of installing a Linux workstation if you did not already have one. For those new to Linux, Chapter 3, Introducing Linux, provided an overview and introduction. In Chapter 4, The host development environment you configured your workstation and networking, and installed some software that we will need, starting with this chapter. In Chapter 5, Eclipse integrated development environment you learned about, and became familiar with, the Eclipse integrated development environment. Chapter 6, The hardware introduced the target single board computer that we will be using throughout this section of the book.

Linux for Embedded and Real-time Applications.
DOI: http://dx.doi.org/10.1016/B978-0-12-811277-9.00007-9

Now it is time to put all that together to build application programs in a cross-development environment.

## ARM I/O Architecture

Like most contemporary processor architectures, the ARM places peripheral devices in the same address space as memory. This means that peripherals can be accessed with any instructions that access memory. The program need not know or care whether it is accessing memory or a hardware device. This is in contrast to, e.g., the x86 architecture, where peripheral devices exist in a separate address space accessed by separate input/output instructions.

In the case of the x86, not only is I/O in a separate address space, the I/O instructions can only be executed from Privilege Level 0, also known as "Kernel space". This necessitates the use of Kernel space device drivers to access peripherals. The ARM, on the other hand, lets us access peripherals directly from a user application. That is not to say that accessing peripherals directly from an application is a good idea, but it is a useful place to start.

The memory map in Table 7.1 shows in abbreviated fashion how the 32-bit address space of the BBB's Sitara processor is allocated. Table 7.1 only shows the address areas that we are interested in. For more complete detail, download the *AM335x and AMIC110 Sitara Processors Technical Reference Manual* from Texas Instruments. See the Resources section for the web address.

**Table 7.1: Sitara SoC memory map**

| Function | Start Address | End Address | Size | Usage |
|---|---|---|---|---|
| External RAM | 0x8000_0000 | 0xBFFF_FFFF | 1 GB | |
| MMC0 | 0x4806_0000 | 0x4806_FFFF | 4 KB | Micro SD card |
| MMC1 | 0x481D_8000 | 0x481D_8FFF | 4 KB | eMMC (4 GB) |
| GPIO0 | 0x44E0_7000 | 0x44E0_7FFF | 4 KB | |
| GPIO1 | 0x4804_C000 | 0x4804_CFFF | 4 KB | |
| GPIO2 | 0x481A_C000 | 0x481A_CFFF | 4 KB | |
| GPIO3 | 0x481A_E000 | 0x481A_EFFF | 4 KB | |
| ADC | 0x44E0_D000 | 0x44E0_EFFF | 8 KB | |

There are a couple of interesting things to note about this map. First, it seems that modern ARM SoCs tend to put peripheral registers "in the middle" of the address space, starting around 0x4400_0000 up to about 0x5700_0000. In my previous experience, peripheral registers usually occupy the top few pages of the address space. The address space is divided between two levels of interconnect busses called "L3" and "L4". Memory and high speed peripherals are attached to L3, and lower speed peripherals are attached to L4. Much of the address space is marked as "Reserved".

### LEDs

The four user LEDs use bits 21 to 24 of GPIO1 based at address 0x4804C000. Fig. 7.1 shows how the LEDs are mapped to bits in the GPIO1 data register.

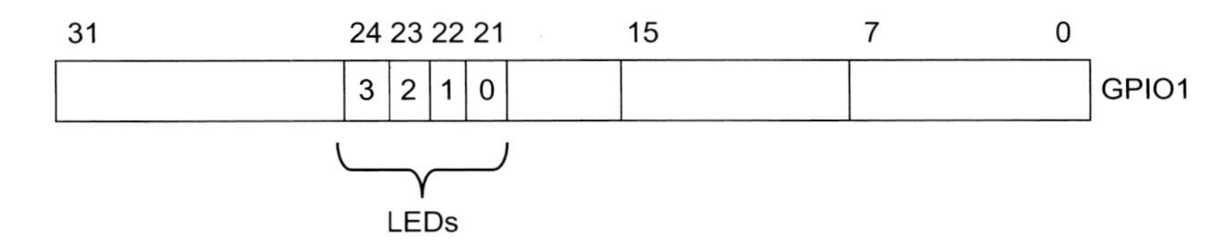


**Figure 7.1**
Layout of LEDs in GPIO1.

The Sitara's GPIO ports are quite complex, with a large number of registers. Each of the 32 bits in a GPIO port can serve as input or output, an interrupt source, or a debounced keypad input. Data output bits may be directly written with 1 or 0, or individually set or cleared by writing a 1 to the corresponding set or clear register. Table 7.2 lists some of the registers contained in each GPIO port. These are the ones that are most relevant to our example right now.

**Table 7.2: GPIO register offsets**

| Offset | Name | Description |
|---|---|---|
| 0x134 | GPIO_OE | Input or output. 1 = input, 0 = output |
| 0x138 | GPIO_DATAIN | Input data register. Returns value of pins. Writing has no effect |
| 0x13C | GPIO_DATAOUT | Output data register. Pins reflect what is written here |
| 0x140 | GPIO_LEVELDETECT0 | Enables interrupt generation on low level (0) detect |
| 0x144 | GPIO_LEVELDETECT1 | Enables interrupt generation on high level (1) detect |
| 0x148 | GPIO_RISINGDETECT | Enables interrupt generation on rising edge, 0 to 1 transition |
| 0x14C | GPIO_FALLINGDETECT | Enables interrupt generation on falling edge, 1 to 0 transition |
| 0x150 | GPIO_DEBOUNCENABLE | Enables the debounce feature |
| 0x154 | GPIO_DEBOUNCINGTIME | Sets the debounce time for any bits in this port that have debounce enabled. Time = value*31usec |
| 0x190 | GPIO_CLEARDATAOUT | Writing 1 clears data out bit to 0. Writing 0 has no effect |
| 0x194 | GPIO_SETDATAOUT | Writing 1 sets data out bit to 1. Writing 0 has no effect |

## Accessing I/O From Linux: Our First Program

### Creating a Project

There are two ways to create a project in Eclipse, a *standard* project, also known as a *Makefile* project, or a *managed* project, also known simply as an *Executable*. A managed project automatically creates a project directory, a template C source file, and a makefile in the default workspace. In Chapter 5, Eclipse integrated development environment we created a managed project for the `record_sort` program.

Typically though, you have already got a number of projects put together in the "conventional" way that you would like to bring into the Eclipse C Development Toolkit (CDT) environment. Most of the sample projects in the book already have makefiles. The role of a Makefile project, then, is to bring an existing makefile, and the files it builds, into the CDT environment.

In the previous chapter you untarred `samples.tar.gz` into your home directory. Now move the resulting `samples/` directory to `/home` in your target file system.

For our first project, we will use the program in `/home/samples/src/led`[1] of your target file system to illustrate some basic concepts and verify that CDT is functioning properly, and that we can execute a newly built program on the target. Then we will look at the `led` program in more detail.

Start up Eclipse and select `File ->New ->C Project` to bring up the New Project wizard as we did in Chapter 5, Eclipse integrated development environment. Enter the project name, "led", and uncheck the `Use default location` box. Now browse to `home>samples>src>led`, and select it by clicking OK. In the `Project type:` window expand `Makefile project`, and select `Empty Project`. Click `Finish`. The project wizard creates the project, which then appears in the Project Explorer window on the left.

There is a very specific reason why the project was created under `home/samples` rather than the default workspace. The final executable must be visible to the target board, so we have to locate it somewhere in the target's Network File System (NFS)-mounted root file system. The alternative would be to create a workspace located in the root file system.

The project should automatically build as part of the creation process. If not, right-click the led entry in the Project Explorer view, and select Build Project. OK, we have built the project, now how do we run it?

### *The Target Execution Environment*

Before proceeding, let us review our set up. The target board has a kernel booted from eMMC flash, and a root file system mounted over the network from the host workstation. `stdin`, `stdout`, and `stderr` on the target are connected to `ttyO0`, which in turn is physically connected to `ttyS0` or `ttyUSB0` on the host. We communicate with the shell running on the target through the terminal emulation program `minicom` Fig. 7.2.

[1] From here on, `/home/samples/` will be left off of path names when the usage is unambiguous.

One consequence of mounting the root file system over NFS is that we can execute *on the target* program files that physically reside on the host's file system. This allows us to test target software without having to program it into flash. And of course, programs running on the target can open files on the NFS-mounted host volume.

One of the directories on the target file system is `/home`. Under `/home/samples/src/` are several subdirectories, each representing a project described later on in the book.

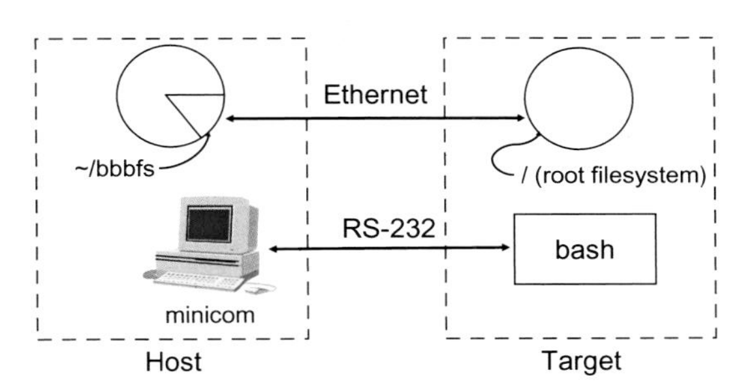


**Figure 7.2**
The host and the target.

Start up `minicom` in a shell window, and boot the target board. When the boot completes, log in as root and execute `ls home/samples` in the `minicom` window. You will see the same files and subdirectories that are in `~/bbbfs/home/samples` on your workstation.

Note that USR0 is already blinking. If you `cat /sys/class/leds/beaglebone\:green\:usr0/trigger` (do not forget TAB autocompletion), you will see that USR0 is being used as a "heartbeat". If you `cat` the trigger file for the other LEDs, you will see that they also have specific triggers. These triggers would interfere with our led program, so we need to turn them off.

`/sys` is another pseudo file system like `/proc` that was described in Chapter 3, Introducing Linux. Like `/proc`, `/sys` is a mechanism to communicate between user space and kernel space. In this particular instance, we can exercise control over system peripherals. Have a look at the script file `samples/src/leds-off.sh`. It writes the string "none" to the `trigger` files of all four LEDs. This says do not trigger the LEDs on anything.

Now, to execute the `led` program, do:

```
cd /home/samples/src/led
./led
```

The program prints a greeting message and then sequentially flashes the four LEDs. Incidentally, the "./" notation in the command represents the current directory. The reason for that is that the current directory is not normally part of a normal user's path. To do so is considered a security risk.

Note that Linux changes the way we have traditionally done embedded programming, as shown in Fig. 7.3. Embedded systems almost always comprise some combination of RAM and nonvolatile memory in the form of ROM, PROM, EPROM, or flash. The traditional way to build an embedded system is to create an executable image of your program, including all library functions statically linked, and perhaps a multitasking kernel. You then load or "burn" this image into one or more nonvolatile memory chips. When the system boots, the processor begins executing this image directly out of ROM.

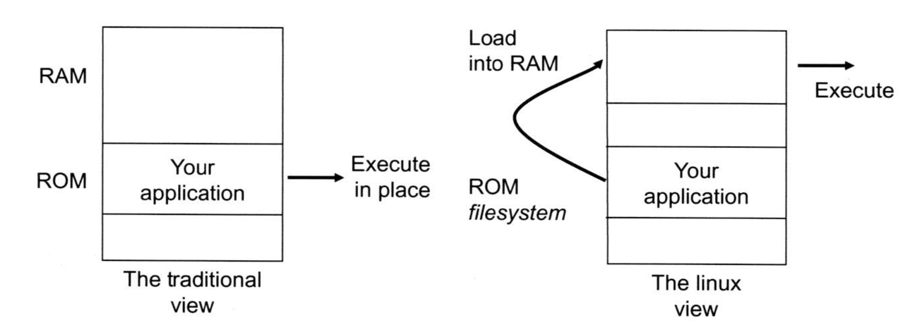


**Figure 7.3**
Two views of embedded programming.

In the Linux view, programs are "files" that must be loaded into memory before executing. Therefore, we create a ROM "file system" containing file images of whatever programs the system needs to run. This may include various utilities and daemons for things like networking. These programs are then loaded into RAM by the boot initialization process, or as needed, and they execute from there.

Generally, the C library is not statically linked to these image files, but is dynamically linked so that a single copy of the library can be shared by whatever programs are in memory at a given time.

One of the advantages of the Linux approach is that we are not confined to loading program files from the ROM file system. As just demonstrated, we can just as easily load programs over a network for testing purposes.

## The `led` Program

Let us take a closer look at the led program as a way of understanding how to access peripherals from user space in Linux. Open `led.c` with the Eclipse editor. You will probably want to turn on line numbers in the editor. Right-click in the marker bar and select `Show Line Numbers`.

Note first of all the "?" symbols next to all of the `#include` directives. If you mouse over one of these, it says "Unresolved inclusion". This means that Eclipse could not find the header files, and thus cannot itself resolve the symbols in the header files. The project will, in fact, build even though Eclipse reports errors for all the symbols defined by the header files. Also, you cannot directly open header files from the Outline view.

This appears to be an artifact of makefile projects using a cross-compiler. The managed `record_sort` project does not have this problem. Here is how to fix it:

1. Right-click the project entry in the `Project Explorer` view, and select `Properties` down at the bottom of the context menu.
2. Expand the `C/C++ General` entry in the `Properties` dialog, and select `Paths and Symbols`.
3. Make sure the Includes tab is selected. Click `Add...` and enter `/usr/local/arm-unknown-linux-gnueabi/arm-unknown-linux-gnueabi/sysroot/usr/include`. Check `Add to all configurations` and `Add to all languages`.
4. Click `Add...` again, and enter `../../include`. This picks up the local header file `am3358-regs.h` that defines peripheral registers. Check `Add to all configurations` and `Add to all languages`.
5. Click `OK`. Click `Apply`. You will be asked if you would like to rebuild the index. Yes, you would.
6. Click `OK` one more time. Fig. 7.4 shows the final result.

The "?" symbols will magically disappear. Note that this is a project-level setting. There does not appear to be a global setting, and that probably makes sense. As we will see later on in this chapter, there is a way to import the settings we just changed into new projects.

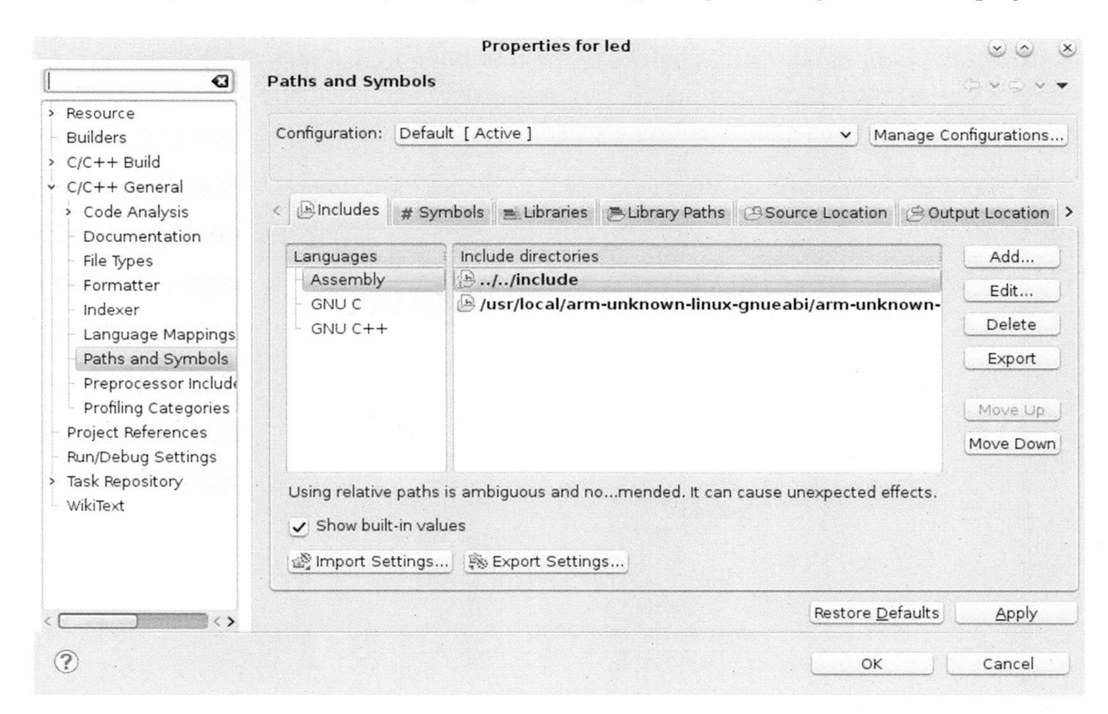


**Figure 7.4**
Adding include paths to a project.

Back to the program. On or about line 37 in `main()`, `GPIOp` is declared to be of type `AM3358P_GPIO`. The latter is a pointer to a structure declared in `am3358-regs.h` that maps the registers for GPIO1 in the address space.

Around line 42, the program opens device `/dev/mem`. The `mem` device is a way of referencing memory directly from a user space program. Three lines later, we call `mmap()` to map the GPIO1 register section beginning at 0x4804C000 (GPIO1) into our process context. Both read and write access to the mapped space are allowed, and it is declared as *shared*, meaning that other processes can simultaneously map to the same space. In your minicom window, do `ls –l /dev/mem` and note that it is only writable by root. That is why you need to log into the BBB as root, in order to run the `led` program.

If `mmap()` succeeds, it returns a pointer, a virtual address, that represents the specified physical space in this process context. Normally, we would have to configure the GPIO1 bits representing the LEDs as outputs, but the system has already done that for us.

Another useful program is `devmem.c` in the directory `src/devmem/`. It maps a specified region of memory, and then lets you read and write locations in that region as bytes, words, or long words. Have a look. The code is fairly self-explanatory.

### The Makefile

It is worth taking a look at the project's makefile for a couple of reasons. Open the makefile in the Eclipse editor (Listing 7.1). The first thing to notice is the definition of CC. There are several related conventions for unambiguously naming GNU tool chains. Shown here is the three-part convention consisting of:

```
<processor>-<operating system>-<tool>
```

```
#
#
# Makefile for ARM/Linux user application
#

SHELL = /bin/sh

CC = arm-linux-gcc

all: led

led: led.c
	${CC}-I../../include -g-o $@ $<

clean:
	rm-f *.o
	rm-f led
```

**Listing 7.1**
Makefile for led project.

Go to `/usr/local/arm-unknown-linux-gnueabi/bin` and you will find a whole set of these three-part file names that are links to corresponding five-part file names consisting of:

```
<processor>-<manufacturer>-<operating system>-<executable binary format>-<tool>
```

In this case, which turns out to be common these days, the manufacturer field is set to "unknown" or "none". I've also seen a format that includes the version of the C library.

## A Data Acquisition Example

Let us move on to something a little more interesting. The Sitara SoC incorporates an 8-channel, 12-bit A/D converter. This can be configured as eight general purpose analog channels, or a mixture of resistive touchscreen inputs and general purpose analog. We will use channel 0 as the basis of a simple data acquisition example that will serve throughout the remainder of this section of the book.

### Set Up the Hardware

Time to get out the ole' soldering iron. In the previous chapter, I suggested you purchase a prototyping cape and a pair of expansion headers. Now is the time to attach the expansion headers to the board. Then you will add a potentiometer (pot), as shown in the schematic of Fig. 7.5. I chose 10 kohms to minimize the current drawn.

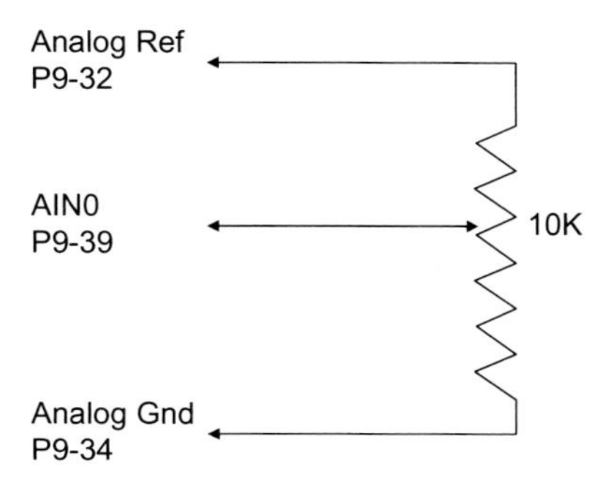


**Figure 7.5**
Analog input schematic.

The analog inputs are limited to 1.8 volts, which is the value of the AD REF pin on P9. So connect the positive end of the pot to P9-32. Fig. 7.6 is a photo of my prototype cape.

In order to plug a cape onto the BBB, you have to remove the debug serial port. Oddly, my USB serial port is no longer working, so I logged into the board using SSH. Log in as root:

```
ssh root@192.168.1.147
```

Substitute your board's actual IP address.

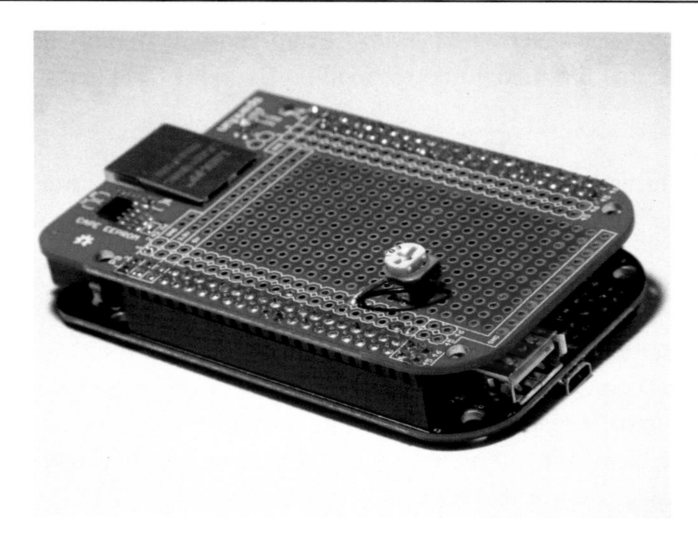

**Figure 7.6**
Prototype cape with potentiometer.

### *Accessing the Analog Subsystem*

The analog subsystem is fairly complex, and we won't try to access it directly from user space. Instead we will use the /sys file system, similar to what we did earlier to turn off the LED triggers. The Resources section references a couple of pages with "recipes", if you will, for setting up the `/sys` files to read the ADC. This completely nonintuitive procedure is embodied in the script file `samples/src/adc-on.sh`. The first line essentially installs the ADC subsystem and creates a set of files under `/sys/devices/ocp.x/helper.y`, named `AIN0` to `AIN7`. The second line shows you where these files actually are. Once you have determined the values for *x* and *y*, you can eliminate the second line. They will not change.

To read an analog input, you just open the corresponding `AIN` file and read it. These files return the value in millivolts, so the range is 0 to about 1800. There is a second set of files in `/sys/devices/ocp.x/ 44e0d000.tscadc/tiadc/iio:device0` called `in_voltage0_raw` to `in_voltage7_raw` that return the raw 12-bit value.

Good programming practice dictates that there be a clear separation between code that is hardware independent, and code that actually manipulates I/O registers. At the very least, hardware-dependent code should be in a separate module from the purely application code. In the LED example, we "cheated" somewhat, in that the hardware-dependent code was mixed in with the main program in a single source file. In that case it did not matter much, because the objective was to illustrate how hardware access works without introducing any kind of application.

In this example we will make a clear distinction between hardware-dependent code and the main program's algorithm. Create a new Eclipse standard C Makefile project named `measure` located at `src/measure`. Open the source file `measure.c`. The first thing to notice is that once again we have those annoying "?" symbols next to the header files. But rather than laboriously retyping the paths to the headers, this time we will *export* some project settings from the `led` project, and *import* them into `measure`.

Proceed as follows:

1. Select the `led` project entry in the `Project Explorer` view. Right click and select `Properties`.
2. Expand `C/C++ General`, and select `Paths and Symbols`. Click the `Export Settings...` button.
3. Make sure the `led` project is selected, and uncheck `# Symbols`.
4. Browse to `home/samples/src/`, and enter a file name. I used `includes`. The data will be saved as an XML file.
5. Fig. 7.7 shows the final export dialog. Click Finish.

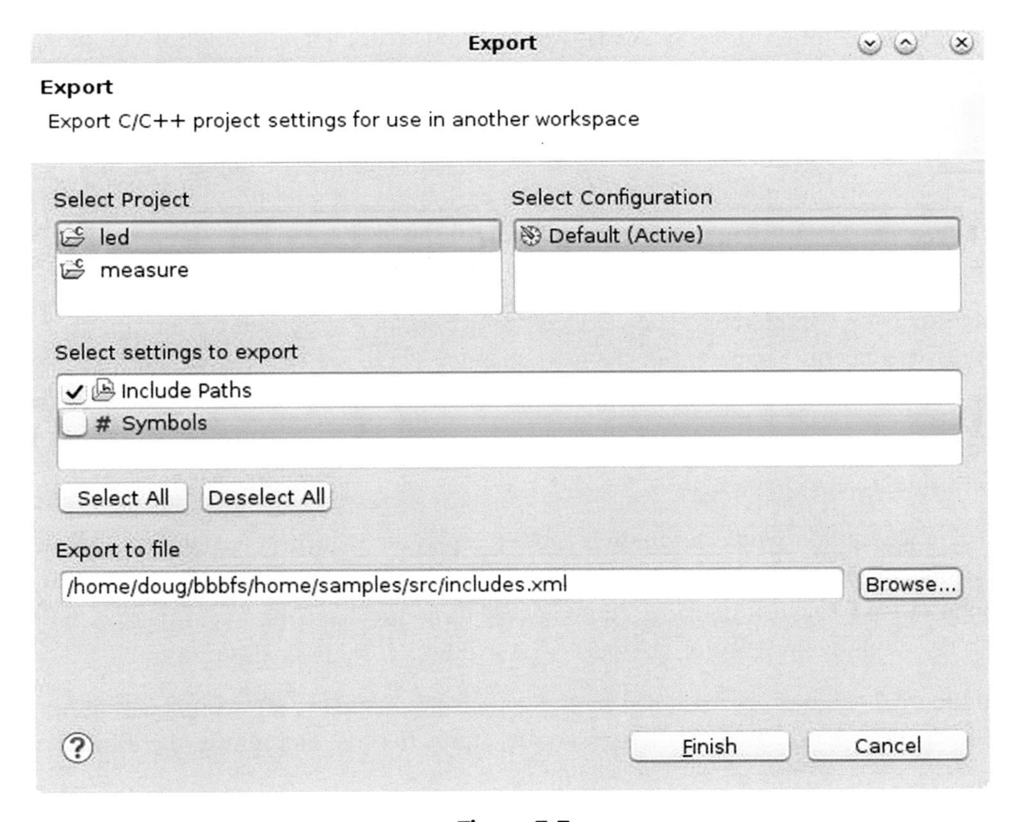


**Figure 7.7**
Export dialog.

Next are the steps for importing the project settings into the measure project:

1. Right-click the `measure` project entry in Project Explorer, and select Properties.
2. Expand C/C++ General and select `Paths and Symbols`. Click the `Import Settings...` button.
3. Browse to `home/samples/src/`, and select `includes.xml`. Click OK.
4. Click Finish.
5. Open the `measure` project properties, and verify that the include paths are present.

The program reads the A/D, and prints the value on `stdout`. It takes a command line argument that is the number of seconds between readings. The default is two.

This is a good time for me to make an observation about my sample programs. There is very little in the way of error or "sanity" checking in these examples. If you enter something incorrectly, chances are the program will crash. That is not much of a problem in this example, but can be a problem in later examples that require more extensive input.

I feel that adding a lot of error checking code gets in the way of communicating the point the example is trying to make. Feel free to add some error checking logic if you like.

Back in the program, the important point to note is that most of the details of dealing with the A/D converter and digital I/O are "hidden" behind APIs (Application Programming Interface) that present data in a form useful to the application. The application need not concern itself with the details of setting up the peripherals, it simply calls the appropriate "init" functions. Then it reads and writes data until the user terminates the program. Finally, it "closes" the peripheral devices.

The file `trgdrive.c` implements the "device driver" APIs. The reason for naming it "trgdrive" will become apparent in the next chapter. There are two peripheral initialization functions: `initAD()` at line 19 and `initDigIO()` at line 54.

`initDigIO()` opens `brightness` files for two of the user LEDs in `/sys/class/leds`. We will use these in the next chapter.

`initAD()` is where you would normally expect to open a connection to the analog input file. But it turns out that the `open()` call triggers the A/D converter. I would have expected that to happen in `read()`, but it doesn't. There appear to be mechanisms for triggering the A/D through `/sys`, but again it is not obvious. So, for now, `initAD()` is a dummy.

Reading the A/D converter is handled by the generically named `readAD()` function at line 30. It opens the analog channel file, reads one sample, closes the file, and returns the binary value.

Digital output is embodied in three functions: `setDigOut()`, `clearDigOut()`, and `writeDigOut()`. These functions simply write the appropriate text value, "0" or "1", to the corresponding `brightness` file.

`closeAD()` is a dummy. `closeDigIO()` just closes the two `brightness` files.

The Makefile for the `measure` project has several build targets. By default, Eclipse only knows how to build the "all" target if it is present. In the case of the `measure` project, `all` builds something called a "simulation version". We will talk about that in the next chapter. Other build targets require that they be declared to Eclipse. On the right-hand side of the C/C++ perspective is a view normally labeled `Outline`, that shows an outline of whatever is currently displayed in the Editor window. There is another tab in that view labeled `Make Targets` that allows us to specify alternate make targets. Select that tab, right click on the `measure` project, and click `Add Target`.

Fig. 7.8 shows the dialog brought up by `Add Target`. In this case, the `Target Name` is `measure`. The `Make Target` is also measure, so we do not have to retype it. Eclipse provides a shortcut for the common case when the target name and make target are the same. Click the `create` button. Now in the `Make Targets` view, under the measure project, right click on the `measure` target, and select `Build Make Target`. In the `Project Explorer` view on the left-hand side of the perspective, three new files show up: `measure.o`, `trgdrive.o`, and `measure`.

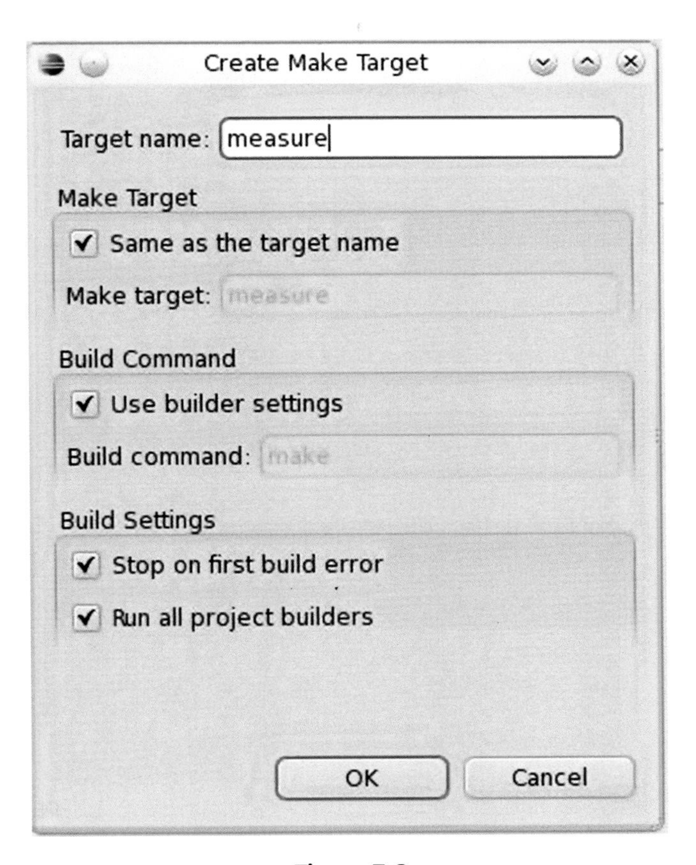


**Figure 7.8**
Create Make Target dialog.

Run the program on the target. Depending on which file you chose to open, the ADC returns either the 12-bit raw data in the range of 0 to 4095, or millivolt data in the range of 0 to 1800. Turn the pot with a small screwdriver, and you should see the reported value change appropriately.

In this chapter, we learned how to access the peripheral registers on the BBB target board, and a little about how to manage peripherals through the `/sys` file system. The next chapter will delve more deeply into the subject of debugging.

## Resources

www.ti.com/lit/pdf/spruh73 – AM335x and AMIC110 Sitara Processors Technical Reference Manual
github.com/CircuitCo/BeagleBone-Black/blob/master/BBB_SRM.pdf – BBB System Reference Manual
github.com/CircuitCo/BeagleBone-Black/blob/master/BBB_SCH.pdf – BBB schematics in PDF
beaglebone.cameon.net/home/reading-the-analog-inputs-adc – Useful write up on accessing the BBB ADC.
https://www.linux.com/learn/how-get-analog-input-beaglebone-black – Another take on the BBB ADC
If you want to investigate the /sys file system in more detail (and you should), check these out:
beagle.s3.amazonaws.com/esc/sysfs-esc-chicago-2010.pdf
www.signal11.us/oss/udev/
www.kernel.org/pub/linux/kernel/people/mochel/doc/papers/ols-2005/mochel.pdf – this paper is a little old, but it is still a good read.

CHAPTER 8

# Debugging embedded software

## Chapter Outline



*If debugging is the process of removing bugs, then programming must be the process of putting them in.*

***Edsger W. Dijkstra***

In Chapter 5, Eclipse integrated development environment you saw how Eclipse integrates very nicely with the Gnu Debugger, GDB. In this chapter, we will look at additional features of GDB, and explore how to use GDB on our target board. We will also consider a simple approach to high-level simulation that can be useful in the early stages of development.

## Remote Debugging With Eclipse

In a typical desktop environment, the target program runs on the same machine as the debugger. But in our embedded environment, Eclipse with GDB runs on the host workstation and the program being debugged runs on the ARM target. You can think of GDB as having both a client and a server side, as illustrated in Fig. 8.1. The client is the user interface, and the server is the actual interaction with the program under test. GDB implements a serial protocol that allows the server and client sides to be separated and communicate either over an RS-232 link or Ethernet.

Linux for Embedded and Real-time Applications.
DOI: http://dx.doi.org/10.1016/B978-0-12-811277-9.00008-0

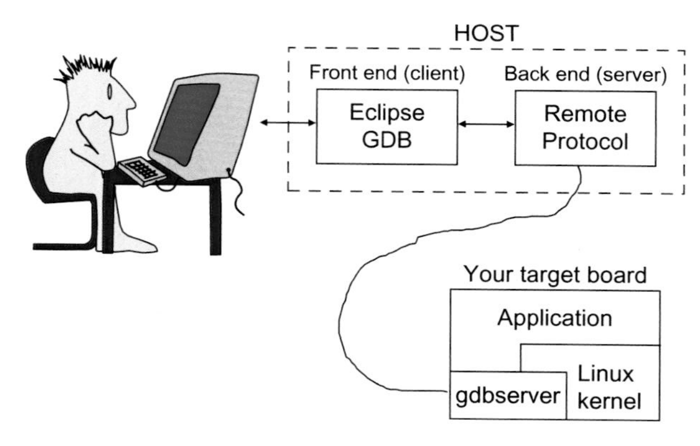


**Figure 8.1**
Client/server nature of GDB.

There are two approaches to interfacing the target to the `gdb` serial protocol:

- `gdb stubs`. A set of functions linked to the target program. `gdb stubs` is an RS-232-only solution.
- `gdbserver`. This is a stand-alone program running on the target that, in turn, runs the program to be debugged. The advantage of `gdbserver` is that it is totally independent of the target program. In other words, the target builds the same regardless of remote debugging. Another advantage of `gdbserver` is that it runs over Ethernet. Finally, Eclipse uses `gdbserver`, so that is the best reason of all to use it.

`gdbserver` is part of the ARM cross-tool chain in `/usr/local/arm-unknown-linux-gnueabi`. It's also already in the BBB's root file system at /usr/bin.

To run the `measure` program under GDB, you would first `cd` to the `src/measure` directory and, as root user, execute:

```
gdbserver :10000 measure
```

The arguments to `gdbserver` are:

- Network port number preceded by a colon. The normal format is "host:port," where "host" is either a host name or an IP address in the form of a dotted quad, but `gdbserver` ignores the host portion. The port number is arbitrary, as long as it does not conflict with another port number in use. Generally, port numbers below 1024 are reserved for established network protocols such as HTTP (port 80), so it is best to pick a number above 1023. Port number 10,000 happens to be the default for the Eclipse debugger.
- Target program name.

- Arguments to the target program, if any.

`gdbserver` responds with something like:

```
Process measure created; pid = 703
```

Your pid value may be different. This says that `gdbserver` has loaded the program, and is waiting for a debug session to begin on the specified port. You must be root user because the program is opening files for writing in /sys that are owned by root.

### *Remote Debug Launch Configuration*

We will need a different debug launch configuration for remote debugging. With the `measure` project selected in the Project Explorer view, bring up the Debug Configuration dialog either from **Run > Debug Configurations**. . ., or click the down arrow next to the cute little bug icon and select Debug Configurations. . . Even though there is a category called C/C++ Remote Application, we are going to create this new configuration under C/C++ Application just like we did for `record_sort`.

1. With C/C++ Application selected, click the **New** button. This brings up the Debug Configurations dialog, as we saw in Chapter 5, Eclipse integrated development environment, with the name and project entries filled out.
2. We need to fill in the C/C++ Application entry. Click **Search Project**. . . That brings up a dialog listing all of the executables found in the project. Select measure and click **OK**.
3. In the center of the dialog is an entry labeled Build (if required) before launching. I recommend selecting Disable auto build. There will be situations later on where auto build would try to build the all target and fail. The failure is innocuous, but it can be disconcerting. Note however that even if you disable auto build, Eclipse may still try to build the all target and output the following message:

   Errors exit in the active configuration of project "measure". Proceed with launch?

   Click Yes. The debug session will proceed correctly.
4. At the bottom of the Main tab is an entry labeled Using GDB Debugger Services Framework (DSF) Create Process Launcher with a link labeled **Select other**. . . Click that link to bring up the dialog shown in Fig. 8.2.
5. Check Use configuration specific settings, and select Legacy Create Process Launcher. Click **OK**.

The issue here is how remote debug sessions are set up and controlled. DSF offers a more seamless, integrated "experience" that we will look at later in the chapter. For now, we will use the simpler approach offered by the Legacy Process Launcher.

**Figure 8.2**
Launcher preference selection.

6. Now select the Debugger tab (Fig. 8.3). Since we are debugging ARM code, we need a version of GDB that understands ARM code. And we have to tell it we are debugging remotely.
7. In the Debugger drop-down, select `gdbserver`.
8. Be sure Stop on startup at main is checked.
9. In the GDB Debugger field enter `arm-linux-gdb`.
10. Select the Connection tab under Debugger Options (Fig. 8.4). For the Type drop-down, select TCP.
11. Enter the IP address of your BBB for the Host name or IP address[1]. Now, click **Apply**.
12. If `gdbserver` is still waiting patiently on the target board, go ahead and click **Debug**. Otherwise, start up `gdbserver` as described earlier before clicking **Debug**.

Eclipse switches (or asks if you want to switch) to the Debug perspective. Even though you are now debugging on the target board, everything looks pretty much like it did when debugging `record_sort` on the workstation.

[1] “Host” is a little misleading here. It really refers to the target board.

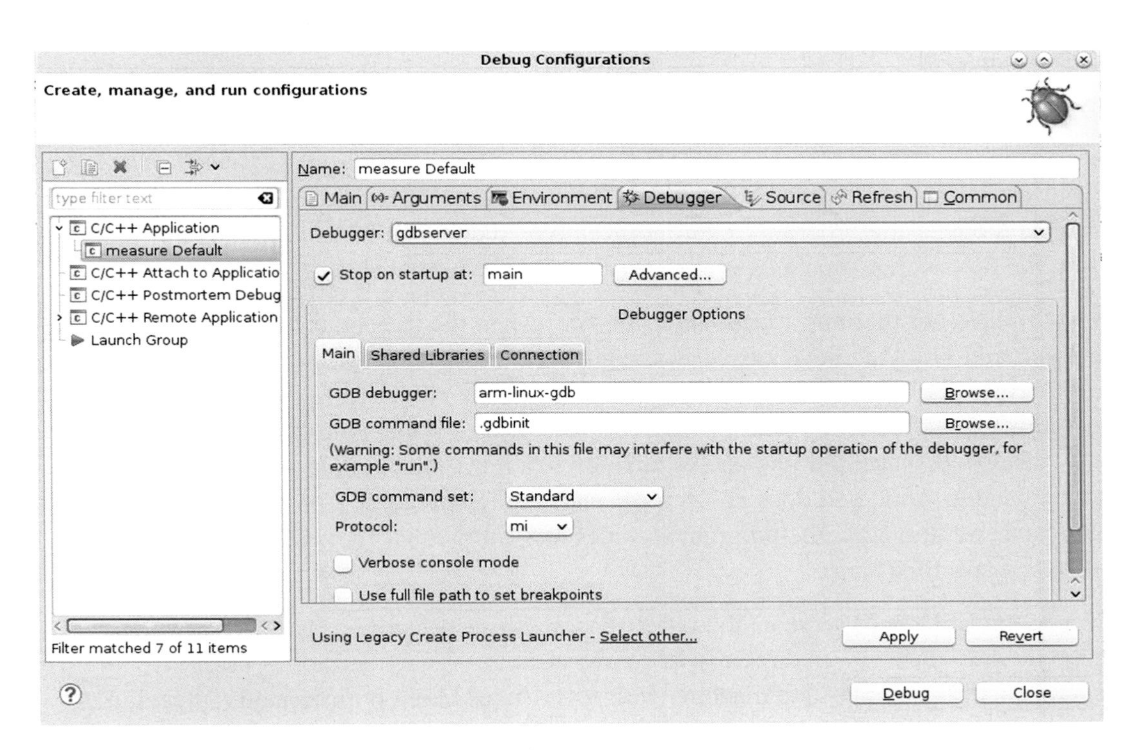


**Figure 8.3**
Debugger tab.

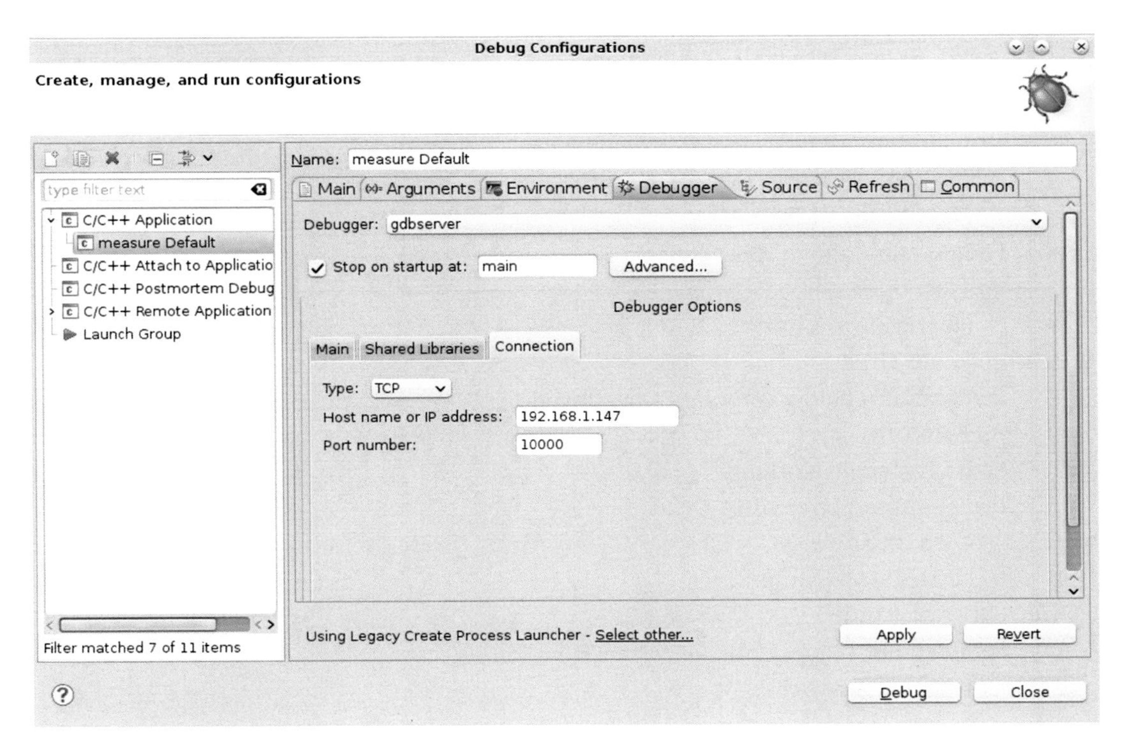


**Figure 8.4**
Connection tab.

## A Thermostat

Let us move on to something a little more interesting, and potentially even useful. We will enhance the measure program by turning it into a "thermostat." We will also take a look at using the host workstation as a preliminary testing environment.

We will have our thermostat activate a "cooler" when the temperature rises above a given setpoint, and turn the cooler off when the temperature falls below the setpoint. In practice, real thermostats incorporate hysteresis that prevents the cooler from rapidly switching on and off when the temperature is right at the setpoint. This is implemented in the form of a "deadband," such that the cooler turns on when the temperature rises above the setpoint + deadband, and does not turn off until the temperature drops below setpoint − deadband. Additionally, the program includes an "alarm" that flashes if the temperature exceeds a specified limit.

Two of the LEDs will serve as the cooler and alarm respectively. These are defined in `driver.h` in the `measure/` directory as `COOLER` and `ALARM`. The thermostat itself is fundamentally a simple state machine with three states based on indicated temperature:

- Normal
- High
- Limit

For each state, the action taken and next state are:

Current state Normal
- Temperature above setpoint + deadband
  - Turn on `COOLER`
  - Next state = High
- Temperature above limit
  - Turn on `ALARM`
  - Next state = Limit

Current state High
- Temperature below setpoint − deadband
  - Turn off `COOLER`
  - Next state = Normal
- Temperature above limit
  - Turn on `ALARM`
  - Next state = Limit

Current state Limit
- Temperature below limit
  - Turn off `ALARM`

Next state = High
Temperature below setpoint − deadband
Turn off `COOLER`
Next state = Normal

The state machine is just a big `switch()` statement on the state variable.

Here is your chance to do some real programming. Make a copy of `measure.c`, and call it `thermostat.c`. Since `thermostat.c` is already a prerequisite in measure's makefile, all you have to do to make it visible to Eclipse is to right click on `measure` in the `Project Explorer` view and select `Refresh`.

Implement the state machine in `thermostat.c`. Declare the variables `setpoint`, `limit`, and `deadband` as global integers. I suggest dividing the value returned by `readAD()` by something in the range of 10 to 20, to get something reasonable as a temperature and to make it easier to adjust the pot. Pick a suitable `setpoint`, and a `limit` another few "degrees" above that. A `deadband` of plus/minus one count is probably sufficient.

## *Host Workstation as Debug Environment*

Although remote GDB gives us a pretty good window into the behavior of a program on the target, there are good reasons why it might be useful to do initial debugging on your host development machine. To begin with, the host is available as soon as a project starts, probably well before any real target hardware is available or working. In many cases, it is easier to accurately exercise limit conditions with a simulation than with the real hardware. The host has a file system that can be used to create test scripts and document test results.

When you characterize the content of most embedded system software, you will usually find that something like 5%, maybe 10%, of the code deals directly with the hardware. The rest of the code is independent of the hardware, and therefore should not need hardware to test it, provided that the code is properly structured to isolate the hardware-dependent elements.

The idea here is to build a simple simulation that stands in for the hardware I/O devices. You can then exercise the application by providing stimulus through the keyboard, and noting the output on the screen. In later stages of testing you may want to substitute a file-based "script driver" for the screen and keyboard to create reproducible test cases.

Take a look at `simdrive.c` in the `measure/` directory. This file exposes the same Application Programming Interface as `trgdrive.c`, but uses a shared memory region to communicate with another process that displays digital outputs on the screen and accepts analog input values via the keyboard. This functionality is implemented in `devices.c`. The shared memory region consists of a data structure of type `shmem_t` (defined in `driver.h`) that includes fields for an analog input and a set of digital outputs that are assumed connected to LEDs. It also includes a process ID field (`pid_t`) set to the pid of the `devices` process that allows the `thermostat` process to signal when a digital output has changed.

`devices` creates and initializes the shared memory region. In `simdrive`, `initAD()` attaches to the previously created shared memory region. `readAD()` simply returns the current value of the `a2d` field in shared memory. The `setDigOut()` and `clearDigOut()` functions modify the `leds` field appropriately, and then signal the `devices` process to update the screen display. Fig. 8.5 illustrates the process graphically.

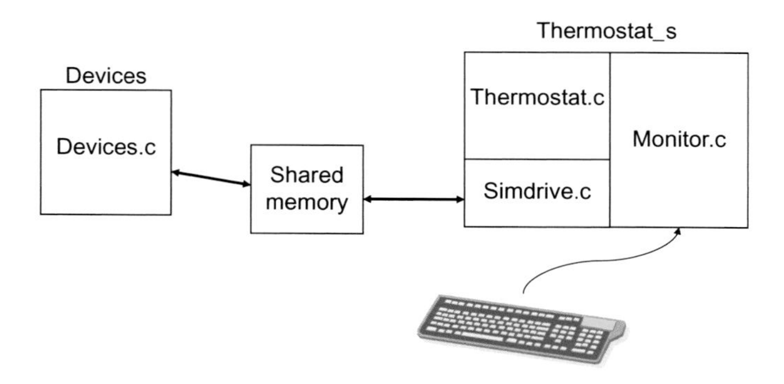


**Figure 8.5**
Thermostat simulation.

The executable for the simulation version of the thermostat is called `thermostat_s` (`_s` for simulation), and is the default target in the project makefile. In fact, it should have been automatically built when you did the Refresh operation to bring `thermostat.c` into Eclipse. To build it again after editing `thermostat.c`, select the `Project` menu and `Build All`.

To build the `devices` program, we need to add another target to the project. Right click on `measure` in `Project Explorer`, and select `Make targets -> Create`. The `Target Name` is "devices," and the `Make Target` is the same. Go ahead and build the devices target.

Run `devices` in a shell window. To debug `thermostat_s`, we can use the same debug launch configuration we created for the `record_sort` project in Chapter 5, Eclipse integrated development environment:

1. Bring up the Debug Configurations dialog, and select the one for the `record_sort` project.
2. Change the name to "host." We will use this configuration for all simulation debugging.
3. Change the project to "measure," and select `thermostat_s` for the application.
4. Click **Apply**, then click **Debug** to bring up the Debug perspective with the program halted at the first executable line of `main()`.

### Advanced Breakpoint Features

The `thermostat_s` program affords us an opportunity to explore some advanced features of breakpoints. Set a breakpoint at the top of the `switch()` statement for your state machine. Now go to the Breakpoints view in the upper right of the Debug perspective, and right-click the breakpoint entry you just created. Select Breakpoint Properties down at the bottom of the context menu. That brings up the rather blank looking dialog of Fig. 8.6.

*Actions* can be attached to a breakpoint such that when it is hit, the attached actions are performed. C Development Toolkit offers five classes of predefined actions:

- **Play Sound** – Play a selected sound file when the breakpoint is hit. Maybe the breakpoint only happens every half hour or so. You can go off and do something else,

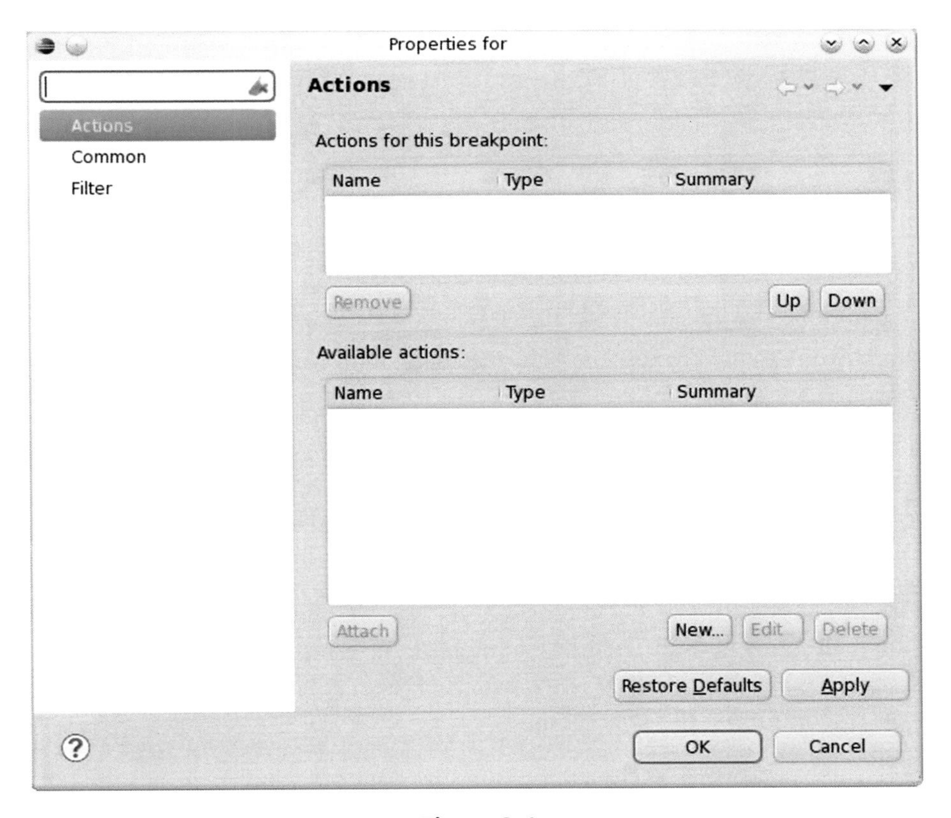


**Figure 8.6**
Breakpoint Properties.

and when you hear the appropriate beep, you can go back and see what happened. Sound files can be `.wav`, `.mid`, `.au`, or `.aiff`.

- **Log Message** – Output a message to the console. To see the message, you must select Log Action Messages in the Console view.
- **Run External Tool** – Execute a program that has been configured in Eclipse as an external tool. For example, the program might be running on a remote device. You could configure the breakpoint to send an e-mail or SMS to your desktop. Programs are installed and configured from `Run>External Tools>External Tools Configurations...`
- **Resume** – Automatically resume the program after a specified delay. Again, if the program is running remotely, this is probably the only way to keep it running after a breakpoint.
- **Reverse Debugging** – This one is new and not described in the Eclipse documentation. There are three possible values: Enable, Disable, and Toggle.

Let's create a sound action. In CentOS 7 most sound files, `.wav`, are found in subfolders of `/usr/share/`. Several are in `sounds/alsa/`. Many others are under `kde4/apps/korganizer/sounds/`. Here is how we do it:

1. In the Actions dialog, click **New**. Sound Action is selected by default.
2. Give it a name. How about "beep"?
3. Browse to one of the folders that contains `.wav` files, and pick one. To hear what it sounds like, click Play Sound.
4. Click **OK**.

The Beep action now shows up in the list of available actions. In a similar fashion, create a Log action. Have it print the message "Hit the breakpoint at the switch statement."

Select each of the entries in the Available actions list, and click **Attach**. They both now appear in the list of Actions for this breakpoint. Let the program run. You should hear the sound when it hits the breakpoint. In the Console view, find the Display Selected Console dropdown near the right hand end of the Console menu. Select Log Action Messages. To return to program output, select `[C/C++ Application] thermostat_s`.

Bring up the Breakpoint Properties dialog again. Select Common in the left-hand column to bring up the dialog in Fig. 8.7. This lists some basic information about the breakpoint, such as the file and line number where it is located, and its enabled status. It also offers us a couple ways of controlling the breakpoint.

Condition lets us specify a Boolean expression that controls whether or not the breakpoint is taken. If the expression evaluates to true, the breakpoint is taken. Otherwise it is ignored. The expression can include any program variables. For example, we are only really interested in the breakpoint on the `switch()` statement when we expect the state to change.

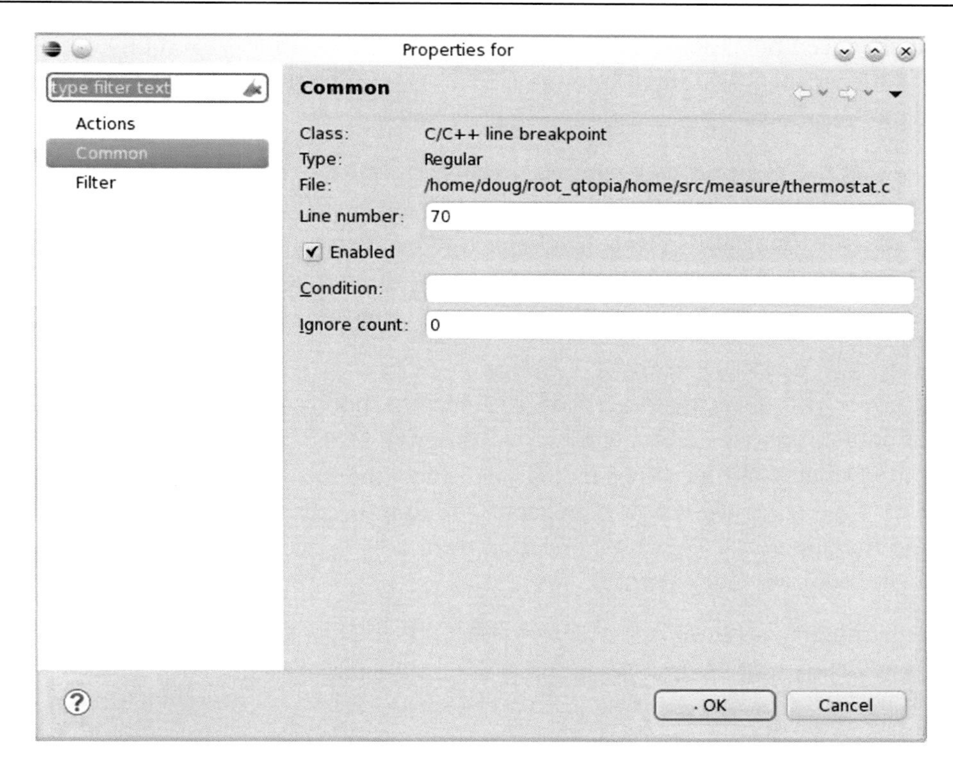


**Figure 8.7**
Common breakpoint properties.

To take the breakpoint on every pass through the loop would be tedious and unproductive. So we could enter the condition:

```
value>setpoint+deadband
```

Then when the breakpoint is taken, we could step through the code to be sure the state is changed and the cooler is turned on. Then we could change the condition to:

```
value>limit
```

and watch what happens at that transition. Then, of course, we would reverse the process and watch the state transitions back to normal.

The Ignore count in the Common dialog is the number of times the breakpoint will be ignored before it is taken. For example, if we enter 10, then the first 10 times the breakpoint is encountered, it will be ignored, and will finally be taken on the 11th encounter. This could be useful in a `for` loop that runs a fixed number of times where we are probably only interested in the first pass through the loop and the last.

Use these advanced breakpoint features to debug your thermostat program by entering different values for the A/D input into the `devices` program, and watch how the thermostat state machine responds.

When you feel the program is running correctly, rebuild it for the target board. You will need another make target. Call it, e.g., `ARMthermo`. Uncheck `Same as the target name`, and enter `all` for the Make target. Uncheck Use builder settings, and enter `make TARGET=1` for the Build command. Before making this target, delete `thermostat.o`. Otherwise, it will not be recompiled for the ARM, and the linker will complain about a bad object file.

To debug `thermostat_t` on the target, you will need to modify the debug launch configuration that you created for the `measure` program at the beginning of the chapter. Bring up the Debug Configurations dialog, and select the measure configuration. Change the name to "target," as we will use this configuration for all of our target debugging. Change the file name in C/C++ Application from `measure` to `thermostat_t`. Those are the only changes required. Click **Apply**.

On the target board, run `gdbserver :10000 thermostat_t`. Then back in Eclipse, click **Debug**. You are now debugging the thermostat program on the target board.

## Debugger Services Framework

For all its capability, the remote debugging process that we have seen so far is a bit tedious. You have to manually start `gdbserver` through the `minicom` shell, and that is where the program console is. Wouldn't it be nice to have `gdbserver` start up automatically when you click **Debug**? And have the program I/O show up in an Eclipse console view. That is what the DSF does for us.

DSF is built on top of the Remote System Explorer (RSE), a collection of tools that allows you to work directly with resources such as files and folders on remote systems. RSE in turn uses SSH, the Secure SHell, to establish communication with remote systems.

### Configuring Remote System Explorer

Fire up Eclipse, and open the RSE perspective. You will find that the Project Explorer view on the left has been replaced by a view labeled Remote Systems, and the collection of views under the Editor has been replaced by something called Remote System Details. Right now, the only thing listed in Remote Systems is Local, your workstation. Expand that entry to reveal Local Files and Local Shells. Expand Local Files to browse the file system (Fig. 8.8).

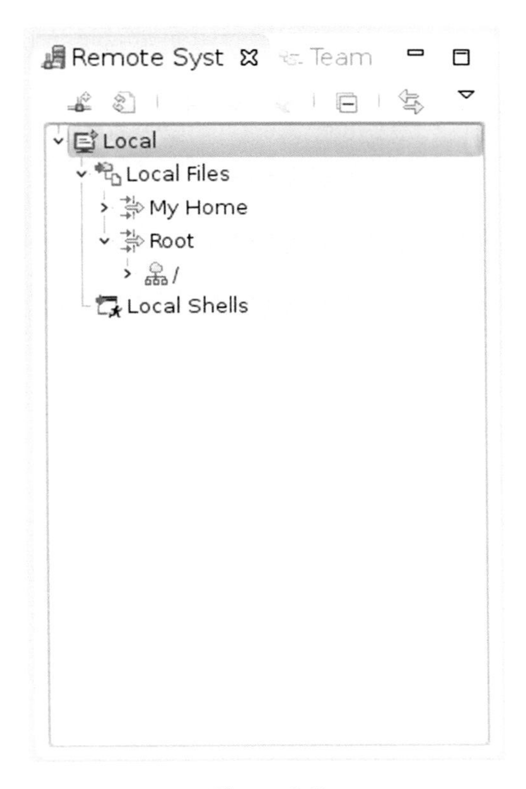


**Figure 8.8**
Remote Systems view.

Right-click on Local Shells, and select Launch Shell. This brings up a Remote Shell view in the lower tabbed window. Commands are entered in the small window at the bottom, and the results show up in the larger window. Fig. 8.9 shows the last few lines of `ls root_qtopia` from the home directory on my system. I find the presentation of this shell a little awkward

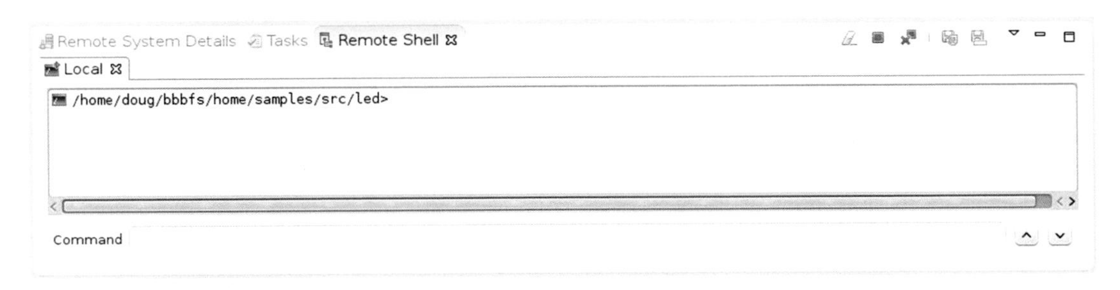


**Figure 8.9**
Local shell view.

and hard to use. There is a more useful shell format available in the remote shell, once we get it set up.

Needless to say, the RSE is not very interesting until we connect it to a remote system, specifically the BBB target board. In the Remote Systems view, click the **Define a connection** icon in the upper left-hand corner, or right-click `Local` and select **New -> Connection**. The first step is to select the Remote System Type. In our case it will be SSH Only.

Why not the Linux type, you may ask? That uses a different, RSE-specific communication protocol called dstore, which has more functionality but requires more resources on the server. dstore is in fact Java-based, so it requires a Java Runtime Engine running on the server, and for the moment that just seems like too much trouble.

Before clicking **Next** to bring up the dialog in Fig. 8.10, be sure your target board is powered up into Linux. Host name defaults to LOCALHOST. Change it to the IP address of your target board. Note that the Connection name is initially the same as the Host name,

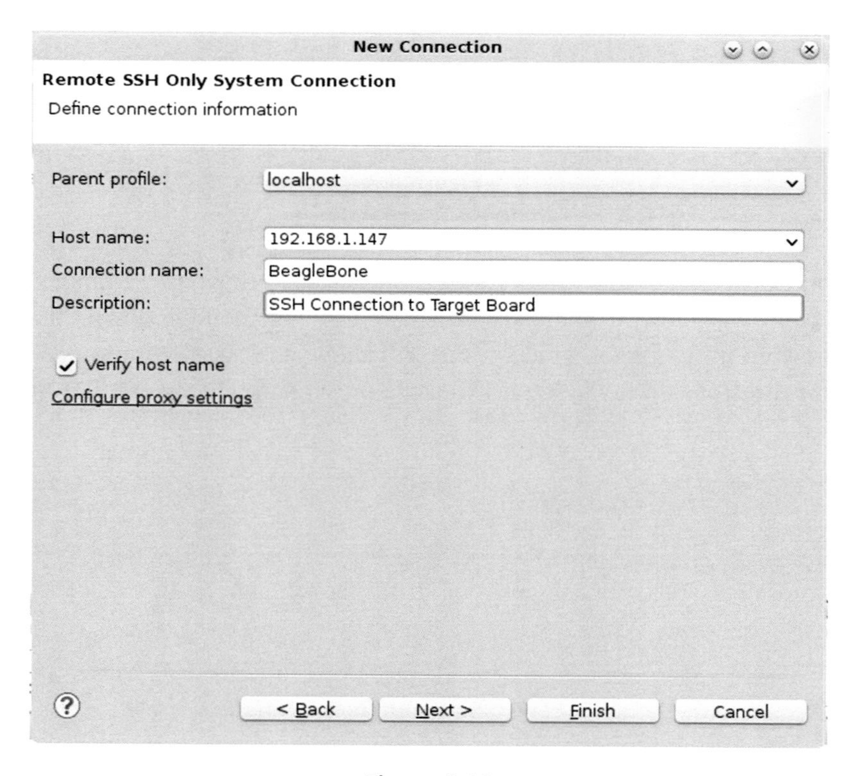


**Figure 8.10**
Setting up an SSH connection.

but you can change it to anything you want, like "BeagleBone" for example. Likewise the Description can be anything.

Clicking **Next** shows the file services available on the remote machine. There is nothing here that needs to be changed. Clicking **Next** again shows the available shell services. Again, nothing needs changing. Clicking **Next** one more time shows the available terminal services. Click **Finish**, and the new connection shows up in the Remote Systems view.

Expand the new connection to reveal entries similar to what we saw with Local. When you expand My Home or Root under Sftp Files, you are required to enter a valid user ID (root) and, optionally, a password for the remote system, which effectively logs you into it. You can now use copy and paste commands to move files between the local host and the remote system. You can also open remote text files in the Eclipse editor by double-clicking them.

Even this is not particularly useful in the present circumstances, where the target's file system is mounted from the workstation over Network File System (NFS). We have direct access to the file system from the workstation. But if the target were truly a remote computer, it would definitely be useful.

Note the entry SSH Terminals under the BeagleBone connection. Right-click that, and select Launch Terminal. This brings up another form of remote shell, as shown in Fig. 8.11. This one looks more like the shell that we have been using through `minicom`, and I personally find it easier to use than the so-called shell view available with the local connection.

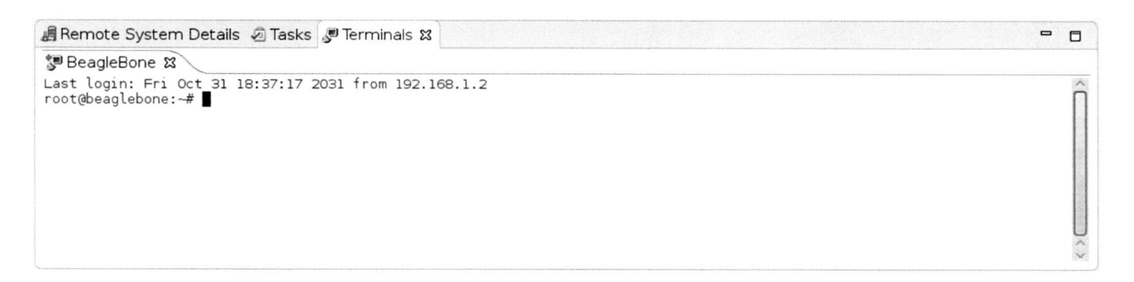


**Figure 8.11**
Ssh Terminal view.

### *Debugging With Remote System Explorer*

We can now set up a debug configuration to use the SSH connection to the target. Bring up the Debug Configurations dialog, select `C/C++ Remote Application`, and click the **New**

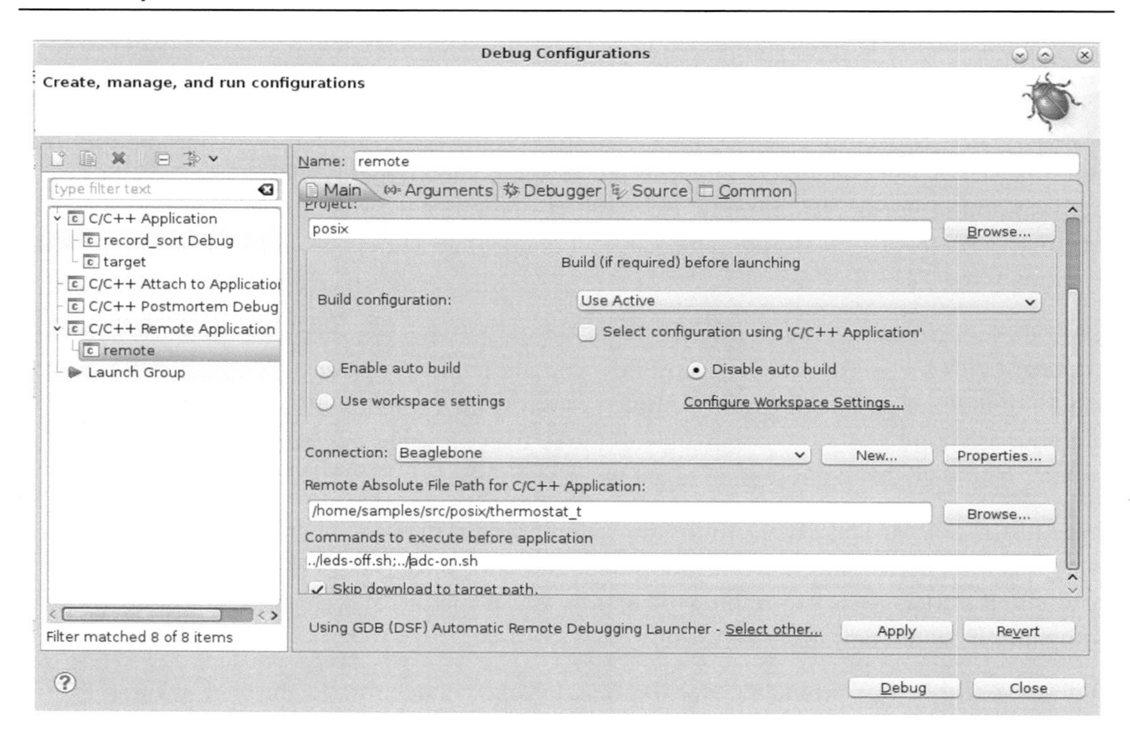


**Figure 8.12**
Remote Debug Configuration.

**launch configuration** icon. Fig. 8.12, shows the final values for this configuration. The top part of the Main tab looks like what we have seen before, selecting the Application and the Project. Again, it is probably best to select `Disable auto build`.

In the `Connection` drop-down, select the remote connection that you just created. Next, we have to tell GDB where to find the executable on the remote target. Enter the path to `thermostat_t` *on the target*. In most cases, RSE would have to "download" the executable in order to execute and debug it on the remote machine. In our case, of course, the executable is already visible to the target in the NFS-mounted file system, so check `Skip download to target path`. If you do choose to let RSE download the executable, do not use the current location, because it will try to overwrite itself and fail. Try `/home/samples/src`. The `Commands to execute before application` entry might be a good place to put the scripts that turn off the LED triggers and enable the ADC subsystem.

Go to the `Debugger` tab, and enter "arm-linux-gdb" for the `GDB debugger`. Select the `Gdbserver Settings` tab, and set the port to 10,000.

Click **Apply** and then **Debug**. After some delay, the Debug perspective will appear, ready to debug `thermostat_t` on the target board. Debugging proceeds just as before, but now the program output appears in an Eclipse console view.

In the next chapter, we will look at multithreaded programming as an alternative to the "heavyweight" Linux process model.

## Resources

openssh.org To learn more about the Secure Shell and specifically the Open Source version.
openssl.org To learn more about the Secure Sockets Layer.

CHAPTER 9

# Posix threads

## Chapter Outline



*Linux is only free if your time has no value.*

***Jamie Zawinski***

The thermostat that we developed in the last chapter is not very practical, because the operating parameters, `setpoint`, `limit`, and `deadband` are hardcoded into the program. Any practical thermostat would make these parameters user adjustable. In this case, we might invent a simple command protocol to change the parameters through the console.

How might we implement such a process? Back in the old DOS days, we might have used the function `kbhit()` to poll the keyboard for activity. But Linux is a multitasking system. Polling is tacky. What we need is an independent thread of execution that patiently waits for a line of text on the console, parses the line, and acts on it.

We could use `fork()` to create a new process. Listing 9.1 illustrates in pseudo code form a possible implementation. We start by creating a shared memory space, and then `fork()` to create a new child process. The child process monitors `stdin`, and posts the result of any valid command to the appropriate entry in the shared memory region. The parent is the `thermostat` program, as we've already seen it.

Linux for Embedded and Real-time Applications.
DOI: http://dx.doi.org/10.1016/B978-0-12-811277-9.00009-2

```
#include <unistd.h>
#include "measure.h"

int running = 1;
params_t *p;    //pointer to shared memory

int main (int argc, void **argp)
{
        create shared memory space;
        switch (fork())
        {
            case -1:
                printf ("fork failed\n");
                break;

            case 0:        // child process
                attach to shared memory space;
                while (running)
                {
                    fgets();
                    parse command;
                    put result in shared memory;
                }
                break;

            default:       // parent process
                attach to shared memory space;
                while (running)
                {
                    read A/D;
                    act on current state;
                }
                break;
        }
        exit (0);
}
```

**Listing 9.1**
Fork implementation of thermostat.

But there is probably a more efficient implementation. Remember from Chapter 3, Introducing linux that a Linux process can have multiple threads of execution, and all of these threads share the process's memory space. So how about we create a thread that just sits on `stdin` waiting for a command? Specifically, in this chapter, we will explore Posix threads.

Posix, also written POSIX, is an acronym that means Portable Operating System Interface, with an X thrown in for good measure. POSIX represents a collection of standards defining various aspects of a portable operating system based on UNIX. These standards are maintained jointly by the Institute of Electrical and Electronic Engineers (IEEE) and the International Standards Organization. The various documents have been pulled together into a single standard in a collaborative effort between the IEEE and The Open Group (see the Resources section).

In general, Linux conforms to Posix. The command shell, utilities, and system interfaces have all been upgraded over the years to meet Posix requirements. But, in the context of multithreading, we are specifically interested here in the Posix Threads interface known as 1003.1c.

The header file that prototypes the Pthreads Application Programming Interface (API) is `pthread.h`, and it resides in the usual directory for library header files, `/usr/include`, or in the case of our cross-tool chain, `/usr/local/arm-unknown-linux-gnueabi/arm-unknown-linux-gnueabi/sysroot/usr/include`.

## Threads

Real-time operating systems often refer to threads as tasks. They are pretty much the same thing. It is an independent thread of execution embodied in a function. The thread has its own stack, referred to as its *context*.

```
int pthread_create (pthread_t *thread, pthread_attr_t *attr, void *(* start_ routine)
   (void *), void *arg);

void pthread_exit (void *retval);

int pthread_join (pthread_t thread, void **thread_return);

pthread_t pthread_self (void);

int sched_yield (void);
```

A thread is created by calling `pthread_create()` with the following arguments:

- `pthread_t` — A *thread object* that represents or identifies the thread. `pthread_create()` initializes this as necessary.
- Pointer to a thread *attribute* object. Often it is NULL. More on this later.
- Pointer to the *start routine*. The start routine takes a single pointer to void * argument, and returns a pointer to void.
- Argument to be passed to the start routine when it is called.

A thread may terminate by calling `pthread_exit()`, or by simply returning from its start function. The argument to `pthread_exit()` is the start function's return value.

In much the same way that a parent process can wait for a child to complete by calling `waitpid()`, a thread can wait for another thread to complete by calling `pthread_join()`. The arguments to `pthread_join()` are the thread object of the thread to wait on, and a place to store the thread's return value. The calling thread is blocked until the target thread terminates. There is no parent/child relationship among threads as there is with processes, so a thread can join any other thread.

A thread can determine its own ID by calling `pthread_self()`. Finally, a thread can voluntarily yield the processor by calling `sched_yield()`.

Note that most of the functions above return an int value. This reflects the Pthreads approach to error handling. Rather than reporting errors in the global variable `errno`, Pthreads functions report errors through their return value.

### Thread Attributes

POSIX provides an open-ended mechanism for extending the API through the use of *attribute objects*. For each type of Pthreads object there is a corresponding attribute object. This attribute object is effectively an extended argument list to the related object create or initialize function. A pointer to an attribute object is always the second argument to a create function. If this argument is NULL, the create function uses appropriate default values. This also has the effect of keeping the create functions relatively simple, by leaving out a lot of arguments that normally take default values.

An important philosophical point is that all Pthreads objects are considered to be "opaque". Most of them anyway. We will see an exception shortly. This means that you never directly access members of the object itself. All access is through API functions that get and set the member arguments of the object. This allows new arguments to be added to a Pthreads object type by simply defining a corresponding pair of get and set functions for the API. In simple implementations, the get and set functions may be nothing more than a pair of macros that access the corresponding member of the attribute data structure.

```
int pthread_attr_init (pthread_attr_t *attr);
int pthread_attr_destroy (pthread_attr_t *attr);
int pthread_attr_getdetachstate (pthread_attr_t *attr, int *detachstate);
int pthread_attr_setdetachstate (pthread_attr_t *attr, int detachstate);
```

*Scheduling Policy Attributes*

```
int pthread_attr_setschedparam (pthread_attr_t *attr, const struct sched_param *param);
int pthread_attr_getschedparam (const pthread_attr_t *attr, struct sched_param *param);
int pthread_attr_setschedpolicy (pthread_attr_t *attr, int policy);
int pthread_attr_getschedpolicy (const pthread_attr_t *attr, int *policy);
int pthread_attr_setinheritsched (pthread_attr_t *attr, int inherit);
int pthread_attr_getinheritsched (const pthread_attr_t *attr, int *inherit
```

Before an attribute object can be used, it must be initialized. Then, any of the attributes defined for that object may be set or retrieved with the appropriate functions. This must be done before the attribute object is used in a call to `pthread_create()`. If necessary, an attribute object can also be "destroyed". Note that a single attribute object can be used in the creation of multiple threads.

The only required attribute for thread objects is the "detach state". This determines whether or not a thread can be joined when it terminates. The default detach state is `PTHREAD_CREATE_JOINABLE`, meaning that the thread can be joined on termination. The alternative is `PTHREAD_CREATE_DETACHED`, which means the thread cannot be joined.

Joining is useful if you need the thread's return value, or you want to make sure threads terminate in a specific order. Otherwise, it is better to create the thread detached. The resources of a joinable thread cannot be recovered until another thread joins it, whereas a detached thread's resources can be recovered as soon as it terminates.

There are also a number of optional scheduling policy attributes.

## *Synchronization: Mutexes*

As soon as we introduce a second independent thread of execution, we create the potential for resource conflicts. Consider the somewhat contrived example of two threads, each of which wants to print the message "I am Thread *n*" on a single shared printer, as shown in Fig. 9.1. In the absence of any kind of synchronizing mechanism, the result could be something like "II a amm TaTasskk 12".

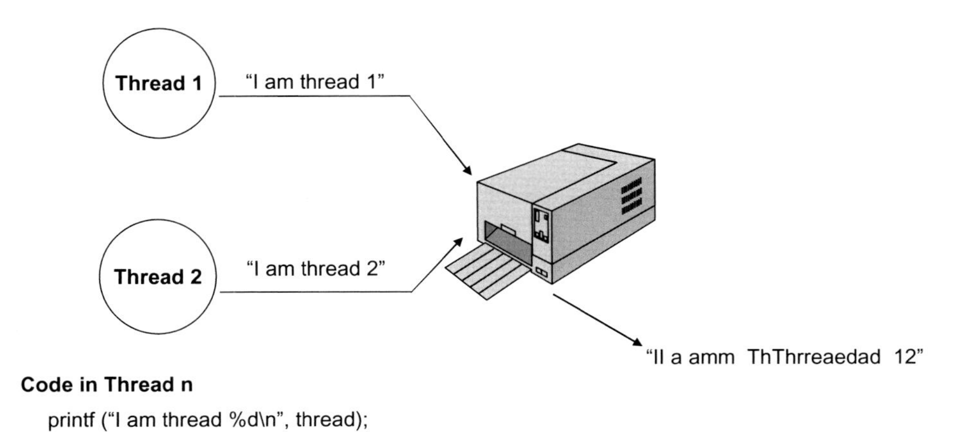


**Figure 9.1**
Resource conflict.

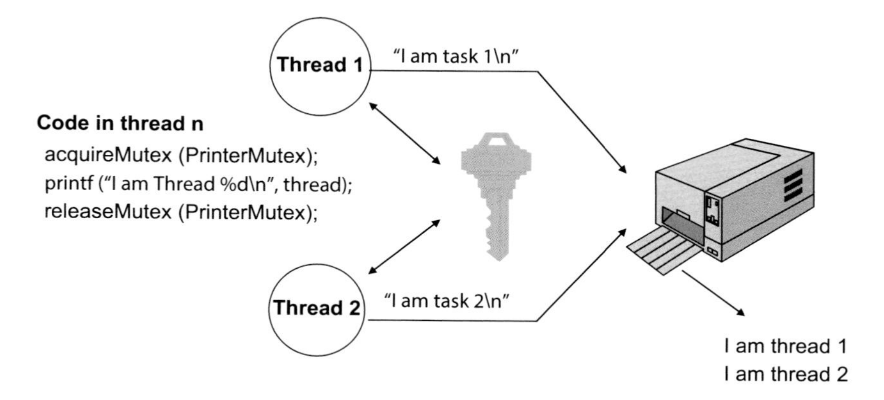


**Figure 9.2**
Solving resource conflict with mutex.

What is needed is some way to regulate access to the printer, so that only one task can use it at a time.

A *mutex* (short for "mutual exclusion") acts like a key to control access to a resource. Only the thread that has the key can use the resource. In order to use the resource (in this case a printer), a thread must first *acquire* the key (mutex) by calling an appropriate kernel service (Fig. 9.2). If the key is available, that is the resource (printer) is not currently in use by someone else, the thread is allowed to proceed. Following its use of the printer, the thread releases the mutex so another thread may use it.

If, however, the printer is in use, the thread is blocked until the thread that currently has the mutex releases it. Any number of threads may try to acquire the mutex while it is in use. All of them will be blocked. The waiting threads are queued, either in order of priority or in the order in which they called `pthread_mutex_lock()`. The choice of how threads are queued at the mutex may be built into the kernel, or it may be a configuration option when the mutex is created.

```
pthread_mutex_t mutex = PTHREAD_MUTEX_INITIALIZER;

int pthread_mutex_init (pthread_mutex_t *mutex, const pthread_mutexattr_t *mutex_attr);
int pthread_mutex_destroy (pthread_mutex_t *mutex);

int pthread_mutex_lock (pthread_mutex_t *mutex);
int pthread_mutex_unlock (pthread_mutex_t *mutex);
int pthread_mutex_trylock (pthread_mutex_t *mutex);
```

The Pthreads mutex API follows much the same pattern as the thread API. There is a pair of functions to initialize and destroy mutex objects, and a set of functions to act on the mutex objects. The listing also shows an alternate way to initialize statically allocated mutex objects. `PTHREAD_MUTEX_INITIALIZER` provides the same default values as `pthread_mutex_init()`.

Two operations may be performed on a mutex: *lock* and *unlock*. The lock operation causes the calling thread to block if the mutex is not available. There is another function called *trylock* that allows you to test the state of a mutex without blocking. If the mutex is available, *trylock* returns success and locks the mutex. If the mutex is not available, it returns `EBUSY`.

### Mutex Attributes

Mutex attributes follow the same basic pattern as thread attributes. There is a pair of functions to create and destroy a mutex attribute object. We will defer discussion of the pshared attribute until later. There are some other attributes we will take up shortly.

The mutex "attribute" type actually started out as a Linux nonportable extension to Pthreads. The Pthreads standard explicitly allows nonportable extensions. The only requirement is that any symbol that is nonportable must have "_np" appended to its name.

Mutex type was subsequently incorporated into the Pthreads standard.

```
int pthread_mutexattr_init (pthread_mutexattr_t *attr);

int pthread_mutexattr_destroy (pthread_mutexattr_t *attr);

int pthread_mutexattr_gettype (pthread_mutexattr_t *attr, int *type);

int pthread_mutexattr_settype (pthread_mutexattr_t *attr, int type);

type =   PTHREAD_MUTEX_NORMAL

            PTHREAD_MUTEX_ERRORCHECK

            PTHREAD_MUTEX_RECURSIVE

            PTHREAD_MUTEX_DEFAULT

int pthread_mutexattr_getprioceiling (const pthread_mutexattr_t *mutex_attr, int
   *prioceiling);
```

(Continued)

**(Continued)**

```
int pthread_mutexattr_setprioceiling (pthread_mutexattr_t *mutex_attr, int
   prioceiling);

int pthread_mutexattr_getprotocol (const pthread_mutexattr_t *mutex_attr, int *protocol);

int pthread_mutexattr_setprotocol (pthread_mutexattr_t *mutex_attr, int protocol);

protocol = PTHREAD_PRIO_NONE

           PTHREAD_PRIO_INHERIT

           PTHREAD_PRIO_PROTECT
```

What happens if a thread should attempt to lock a mutex that it has already locked? Normally the thread would simply hang up. The "type" attribute alters the behavior of a mutex when a thread attempts to lock a mutex that it has already locked. The possible values for type are:

*Normal*. If a thread attempts to lock a mutex it already holds, it is blocked and thus effectively deadlocked. The normal mutex does no consistency or sanity checking.
*Error checking*. If a thread attempts to recursively lock an error checking mutex, the lock function returns immediately with the error code EDEADLK, meaning you cannot do this because you have already locked the mutex. Furthermore, the unlock function returns an error if it is called by a thread other than the current owner of the mutex.
*Recursive*. A recursive mutex allows a thread to successfully lock a mutex multiple times. It counts the number of times the mutex was locked, and requires the same number of calls to the unlock function before the mutex goes to the unlocked state. This type also checks that the mutex is being unlocked by the same thread that locked it.
*Default*. The standard says that this type results in "undefined behavior" if a recursive lock is attempted. It also says that an implementation may map Default to one of the other types.

Optionally, a Pthreads mutex can implement the priority inheritance or priority ceiling protocols to avoid priority inversion. The mutex attribute *protocol* can be set to "none", "priority inheritance", or "priority ceiling", represented by the symbol `PTHREAD_PRIO_PROTECT`. The *prioceiling* attribute sets the value for the priority ceiling.

### *Problems With Solving the Resource Sharing Problem: Priority Inversion*

Using mutexes to resolve resource conflicts can lead to subtle performance problems. Consider the scenario illustrated in Fig. 9.3. Threads 1 and 2 each require access to a common resource protected by a mutex. Thread 1 has the highest priority, and Thread 2 has the lowest. Thread 3, which has no need for the resource, has a "middle" priority.

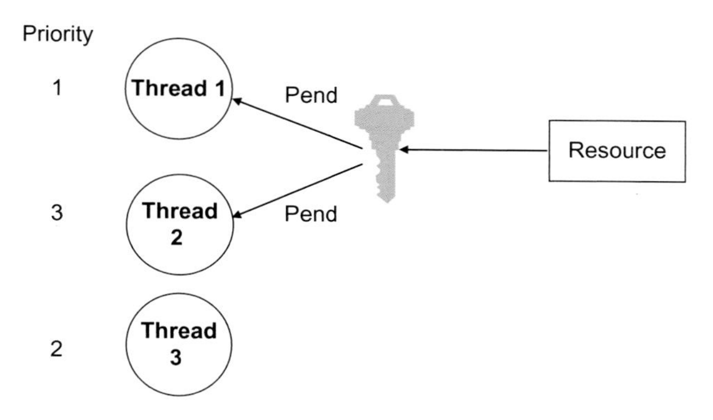


**Figure 9.3**
Priority inversion scenario.

Fig. 9.4 is an execution timeline of this system. Assume Thread 2 is currently executing and pends on the mutex. The resource is free, so Thread 2 gets it. Next, an interrupt occurs that makes Thread 1 ready. Since Thread 1 has higher priority, it preempts Thread 2, and executes until it pends on the `resource` mutex.

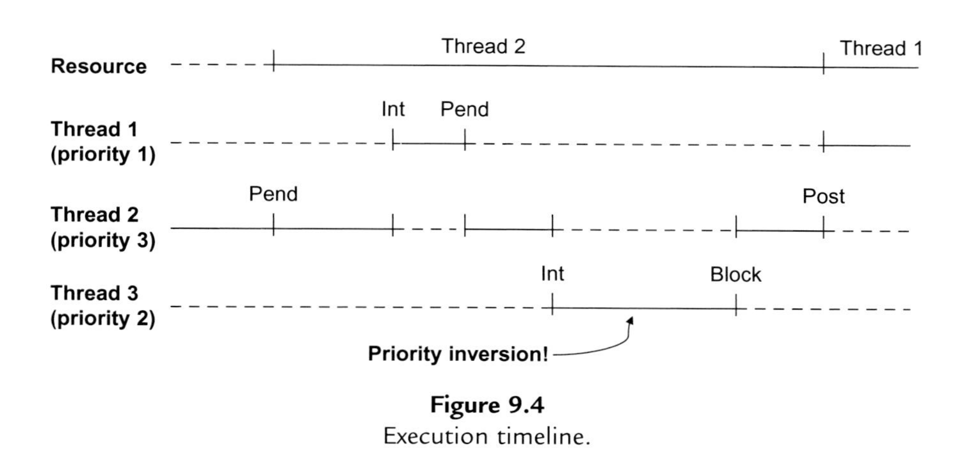


**Figure 9.4**
Execution timeline.

Since the resource is held by Thread 2, Thread 1 blocks, and Thread 2 regains control. So far everything is working as we would expect. Even though Thread 1 has higher priority, it simply has to wait until Thread 2 is finished with the resource.

The problem arises if Thread 3 should become ready while Thread 2 has the resource locked. Thread 3 preempts Thread 2. This situation is called *priority inversion*, because a lower priority thread (Thread 3) is effectively preventing a higher priority thread (Thread 1) from executing.

A common solution to this problem is to temporarily raise the priority of Thread 2 to that of Thread 1 as soon as Thread 1 pends on the mutex. Now, Thread 2 cannot be preempted by anything of lower priority than Thread 1. This is called *priority inheritance.*

Another approach, called *priority ceiling*, raises the priority of Thread 2 to a specified value higher than that of any task that may pend on the mutex as soon as Thread 2 gets the mutex. This is considered to be more efficient, because it eliminates unnecessary context switches. No thread needing the resource can preempt the thread currently holding it.

Posix threads has optional attributes for setting a mutex's protocol as either priority inheritance or priority ceiling, and for setting the priority ceiling value.

## Communication: Condition Variables

There are many situations where one thread needs to notify another thread about a change in status to a shared resource protected by a mutex. Consider the situation in Fig. 9.5, where two threads share access to a queue. Thread 1 reads the queue and Thread 2 writes it. Clearly, each thread requires exclusive access to the queue, and so we protect it with a mutex.

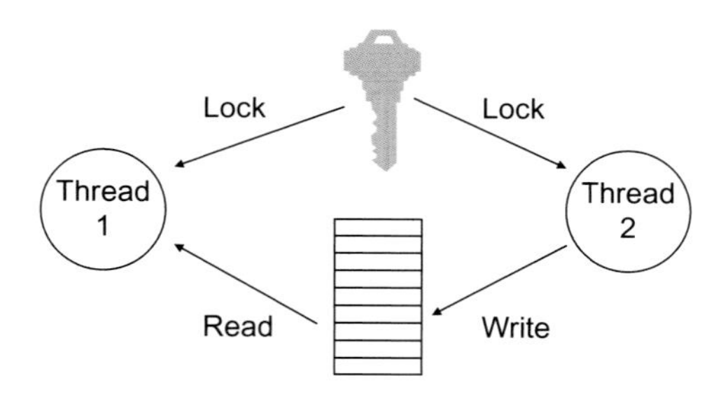


**Figure 9.5**
Communicating via condition variable.

Thread 1 will lock the mutex and then see if the queue has any data. If it does, Thread 1 reads the data, and unlocks the mutex. However, if the queue is empty, Thread 1 needs to block somewhere until Thread 2 writes some data. Thread 1 must unlock the mutex before blocking, or else Thread 2 would not be able to write. But there is a gap between the time Thread 1 unlocks the mutex and blocks. During that time, Thread 2 may execute and not recognize that anyone is blocking on the queue.

The condition variable solves this problem by waiting (blocking) with the mutex locked. Internally, the conditional wait function unlocks the mutex, allowing Thread 2 to proceed. When the conditional wait returns, the mutex is again locked.

```
pthread_cond_t cond = PTHREAD_COND_INITIALIZER;

int pthread_cond_init (pthread_cond_t *cond, const pthread_condattr_t *cond_attr);

int pthread_cond_destroy (pthread_cond_t *cond);

int pthread_cond_wait (pthread_cond_t *cond, pthread_mutex_t *mutex);

int pthread_cond_timedwait (pthread_cond_t *cond, pthread_mutex_t *mutex, const struct
    timespec *abstime);

int pthread_cond_signal (pthread_cond_t *cond);

int pthread_cond_broadcast (pthread_cond_t *cond);
```

The basic operations on a condition variable are *signal* and *wait*. Signal wakes up one of the threads waiting on the condition. The order in which threads wake up is a function of scheduling policy. A thread may also execute a *timed wait*, such that if the specified time interval expires before the condition is signaled, the wait returns with an error. A thread may also *broadcast* a condition. This wakes up all threads waiting on the condition.

### Condition Variable Attributes

Pthreads does not define any required attributes for condition variables, although there is at least one optional attribute.

## Thread Termination and Cancellation

A thread may be terminated either voluntarily or involuntarily. A thread terminates itself either by simply returning, or by calling `pthread_exit()`. In the latter case, all *cleanup handlers* that the thread registered by calls to `pthread_cleanup_push()` are called prior to termination.

Most threads run in an infinite loop. As long as the system is powered up, the thread is running, doing its thing. Some threads start up, do their job, and finish. But there are also circumstances where it is useful to allow one thread to terminate another thread involuntarily. Perhaps the user presses a CANCEL button to stop a long search operation. Maybe the thread is part of a redundant numerical algorithm, and is no longer needed because another thread has solved the problem. The Pthreads cancellation mechanism provides for the orderly shutdown of threads that are no longer needed.

But cancellation must be done with care. You do not just arbitrarily stop a thread at any point in its execution. Suppose it has a mutex locked. If you terminate a thread that has locked a mutex, it can never be unlocked. The thread may have allocated one or more dynamic memory blocks. How does that memory get returned if the thread is terminated?

Pthreads allows each thread to manage its own termination, such that the thread can free up and/or return any global resources before it actually terminates. So, when you cancel a thread you are usually not stopping it immediately, you are asking it to terminate itself as soon as it is safe or convenient.

Pthreads supports three cancellation modes (Fig. 9.6) encoded as two bits called *cancellation state* and *cancellation type*. A thread may choose to disable cancellation because it is performing an operation that must be completed. The default mode is Deferred, meaning that cancellation can only occur at specific points, called *cancellation points*, where the program tests whether the thread has been requested to terminate. Most functions that can block for an unbounded time, such as waiting on a condition variable or reading or writing a file, are cancellation points, and are defined as such in the Posix specification.

| Mode | State | Type | Meaning |
|---|---|---|---|
| Off | Disabled | N/A | Cancellation remains pending until enabled |
| Deferred | Enabled | Deferred | Cancellation occurs at next cancellation point |
| Asynch-ronous | Enabled | Asynch-ronous | Cancellation may occur at any time |

**Figure 9.6**
Cancellation modes.

While asynchronous cancellation mode might seem like a good idea, it is rarely safe to use. That is because, by definition, you do not know what state the thread is in when it gets the cancellation request. It may have just called `pthread_mutex_lock()`. Is the mutex locked? Don't know. So while asynchronous cancellation mode is in effect, you cannot safely acquire any shared resources.

```
int pthread_cancel (pthread_t thread);

int pthread_setcancelstate (int state, int *oldstate);

int pthread_setcanceltype (int type, int *oldtype);

void pthread_testcancel (void);
```

A thread can cancel another thread by calling `pthread_cancel()`. `pthread_setcancelstate()` and `pthread_setcanceltype()` allow a thread to set its cancellation mode. Note that these functions return the previous value of state and type respectively. The function `pthread_testcancel()` allows you to create your own cancellation points. It returns immediately if the thread has not been requested to terminate. Otherwise, it does not return.

### *Cleanup Handlers*

When a thread is requested to terminate itself, there may be some things that need to be "cleaned up" before the thread can safely terminate. It may need to unlock a mutex, or return a dynamically allocated memory buffer, for example. That is the role of *cleanup handlers*. Every thread conceptually has a stack of active cleanup handlers. Handlers are pushed on the stack by `pthread_cleanup_push()`, and executed in reverse order when the thread is cancelled or calls `pthread_exit()`. A cleanup handler takes one `void *` argument.

```
void pthread_cleanup_push (void (*routine)(void *), void *arg);

void pthread_cleanup_pop (int execute);
```

The most recently pushed cleanup handler can be popped off the stack with `pthread_cleanup_pop()` when it is no longer needed. Often, the functionality of a cleanup handler is needed whether or not the thread terminates. The execute argument specifies whether or not a handler is executed when it is popped. A nonzero value means execute. Note also that `pthread_cleanup_ pop()` can only be called from the same function that called `pthread_cleanup_ push()`. Aside from being good programming practice, this is necessary because `pthread_cleanup_push()` is a macro that, in many implementations, ends with an opening brace, "{", introducing a block of code. `pthread_cleanup_pop()` then has the corresponding closing brace.

Listing 9.2 shows the read thread (the main function) of a typical queuing application with a cleanup handler added. We assume the default deferred cancellation mode. Note that `pthread_cleanup_pop()` is used to unlock the mutex, rather than the normal mutex unlock function.

The reason we need a cleanup handler here is that `pthread_cond_wait()` is a cancellation point, and the mutex is locked when we call it. But is it really necessary to push and pop the cleanup handler on every pass through the while loop? It is if there is a cancellation point in the section called "do something with the data" where the mutex is unlocked. This thread can only invoke the cleanup handler if it has the mutex locked. If there are no cancellation points while the mutex is unlocked, then it is safe to move the push cleanup call outside the loop. In that case, we do not really need pop cleanup.

We will have an opportunity to use a cleanup handler in the next chapter.

```
/* Cleanup Handler Example */
#include <pthread.h>

typedef struct my_queue_tag {
   pthread_mutex_t      mutex;          /* Protects access to queue */
   pthread_cond_t       cond;           /* Signals change to queue */
   int                  get, put;       /* Queue pointers */
   unsigned char        empty, full;    /* Status flags */
   int                  q[Q_SIZE]       /* the queue itself         */
} my_queue_t;

my_queue_t data = {
   PTHREAD_MUTEX_INITIALIZER, PTHREAD_COND_INITIALIZER, 0, 0, 1, 0};

void cleanup_handler (void *arg)
/*
   Unlocks the mutex associated with a queue
*/
{
   pthread_mutex_unlock (mutex *) arg);
}

int main (int argc, char *argv[])
{

   while (1)
   {
      pthread_cleanup_push (cleanup_handler, (void *) &data.mutex);
      pthread_mutex_lock (&data.mutex);
      if (queue is empty)
      {
         data.empty = 1;
         pthread_cond_wait (&data.cond, &data.mutex);
      }
      /* read data from queue */
      pthread_cleanup_pop (1);

      /* do something with the data */
   }
}
```

**Listing 9.2**

## *Pthreads Implementations*

Until kernel version 2.6, the most prevalent threads implementation was LinuxThreads. It had been around since about 1996, and by the time development began on the 2.5 kernel it was generally agreed that a new approach was needed to address the limitations in LinuxThreads. Among these limitations, the kernel represents each thread as a separate process, or *schedulable entity*, giving it a unique process ID, even though many threads exist within one process entity. This causes compatibility problems with other thread implementations. There is a hard coded limit of 8192 threads per process, and while this may seem like a lot, there are some problems that can benefit from running thousands of threads.

The result of this new development effort is the Native Posix Threading Library, or NPTL, which is now the standard threading implementation in 2.6 and 3.x series kernels. It too treats each thread as a separately schedulable entity, but takes advantage of improvements in the kernel that were specifically intended to overcome the limitations in LinuxThreads. The `clone()` call was extended to optimize thread creation. There is no longer a limit on the number of threads per process, and the new Completely Fair Scheduler can handle thousands of threads without excessive overhead. A new synchronization mechanism, the Fast Mutex or "futex", handles the noncontention case without a kernel call.

In tests on an IA-32, NPTL is reportedly able to start 100,000 threads in two seconds. By comparison, this test under a kernel without NPTL would have taken around 15 minutes.

## *Upgrading the Thermostat*

We now have enough background to add a thread to our thermostat to monitor the serial port for changes to the parameters. We will use a very simple protocol to set parameters consisting of a letter followed by a space followed by a number. "s" represents the setpoint, "l" the limit, and "d" the deadband. So, to change the setpoint you would enter, for example:

```
s 65<Enter>
```

This sets the setpoint to 65 degrees.

Copy `thermostat.c` that currently resides in `src/measure/` to `src/posix/`. Create a new Eclipse project called "posix", located in the `home/samples/src/posix/` directory. Again, be sure it is a makefile project, and import the project settings with the include paths. Open `monitor.c`. The first thing to notice is #include<pthread.h>. This header file prototypes the Pthreads API. Note the declarations of `paramMutex` and `monitorT`. The latter will become the handle for our monitor thread.

Because the parameters, `setpoint`, `limit`, and`deadband`, are accessed independently by the two threads, the thermostat and monitor, we need a mutex to guarantee exclusive access to one thread or the other. In reality, this particular application probably does not absolutely require the mutex, because the shared resources are simple integers and access to them should be atomic. But it does serve to illustrate the point.

Note, incidentally, that `setpoint`, `limit`, and `deadband` are declared `extern` in `thermostat.h`. It is assumed that these variables are allocated in `thermostat.c`. If you happened to use different names for your parameters, you will need to change them to match.

Take a look at the monitor() function. It is just a simple loop that gets and parses a string from stdin. Note that before monitor() changes any of the parameters, it locks paramMutex. At the end of the switch statement, it unlocks paramMutex. This same approach will be required in your thermostat state machine.

Move down to the function createThread(). This will be called from main() in thermostat.c to initialize paramMutex and create the monitor thread. createThread() returns a nonzero value if it fails. terminateThread() makes sure the monitor thread is properly terminated before terminating main(). It should be called from main() just before closeAD().

### Changes Required in Thermostat.c

Surprisingly little changes to make thermostat.c work with the monitor.

1. include pthread.h and thermostat.h.
2. call createThread(), and test the return value at the top of main().
3. lock paramMutex just before the switch statement on the state variable, and unlock it at the end of the switch statement.
4. call terminateThread() just after the running loop.

That is it! The makefile supports both target and simulation builds, as described in the last chapter. The drivers for both builds come from the measure/ directory, and it is assumed that they are built. Build the simulation and try it out.

## Debugging Multithreaded Programs

Multithreading does introduce some complications to the debugging process but, fortunately, GDB has facilities for dealing with those. Debug thermostat_s in Eclipse (be sure devices is running) using the host debug configuration, and set a breakpoint at the call to createThread() in main(). When it stops at the breakpoint, click **Step into** to go into the createThread() function.

Step over the pthread_mutex_init() call, and CHECK_ERROR. Now, when you step over the call to pthread_create(), a second thread appears in the Debug view. This is the monitor thread, and it is already inside the fgets() call.

Set a breakpoint in the monitor() function just after the call to fgets(). Now, enter a line of text in the Console view. It does not really matter if it is a valid command or not. All we want to do at this point is get the program to stop at the breakpoint. Let the program continue, and when it reaches the breakpoint note that Thread [2] is now the one suspended at a breakpoint. Thread [1] is most likely suspended inside the sleep() function.

Play around with setting breakpoints in various places in both `main()` and `monitor()` to verify that parameters get updated properly and the thermostat responds correctly to the new values.

When you are satisfied with the program running under the simulation, rebuild it for the target and debug it there. You will need to create a new make target, just as we did in the previous chapter for the measure project. You will also need to delete both `monitor.o` and `thermostat.o` for the target version to build properly.

When you start the debugger, you are likely to see some warning messages whiz by in the Console view. These are innocuous.

The debugger stops at the first line of `main()` as normal. Let the program execute to the breakpoint at `createThread()`. Step over, and the Debug view shows two threads just as we saw in the simulation version. From here on out, the target debugging process is essentially identical to the simulation mode.

With a basic understanding of threads, we are ready to tackle networking, the subject of the next chapter.

## Resources

The Open Group has made available free for private use the entire Posix (now called Single Unix) specification. Go to: www.unix.org/online.html

You will be asked to register. Don't worry, it is painless and free. Once you register, you can read the specification on-line, or download the entire set of html files for local access.

This chapter has been little more than a brief introduction to Posix threads. There is a lot more to it. An excellent, more thorough treatment is found in,

Butenhof, David R., *Programming with POSIX Threads*, Addison-Wesley, 1997.

CHAPTER 10

# Embedded networking

## Chapter Outline



*Give a man a fish and you feed him for a day. Teach him to use the Net and he won't bother you for weeks.*

Everything is connected to the Internet, even refrigerators.[1] And of course the big buzz phrase in the technology biz is the "Internet of Things," or IoT. So it is time to turn our attention to network programming in the embedded space. Linux, as a Unix derivative, has extensive support for networking.

[1] http://www.engadget.com/2011/01/12/samsung-wifi-enabled-rf4289-fridge-cools-eats-and-tweets-we-go/ or just Google "internet refrigerator." I don't suppose there are any net-connected toaster ovens yet.

Linux for Embedded and Real-time Applications.
DOI: http://dx.doi.org/10.1016/B978-0-12-811277-9.00010-9

We will begin by looking at the fundamental network interface, the socket. With that as a foundation, we will go on to examine how common application-level network protocols can be used in embedded devices.

## Sockets

The "socket" interface, first introduced in the Berkeley versions of Unix, forms the basis for most network programming in Unix systems. Sockets are a generalization of the Unix file access mechanism that provides an endpoint for communication, either across a network or within a single computer. A socket can also be thought of as an extension of the named pipe concept that explicitly supports a client/server model, wherein multiple clients may be attached to a single server.

The principal difference between file descriptors and sockets is that a file descriptor is bound to a specific file or device when the application calls `open()`, whereas sockets can be created without binding them to a specific destination. The application can choose to supply a destination address each time it uses the socket, for example when sending datagrams, or it can bind the destination to the socket to avoid repeatedly specifying the destination, for example when using TCP.

Both the client and server may exist on the same machine. This simplifies the process of building client/server applications. You can test both ends on the same machine before distributing the application across a network. By convention, network address 127.0.0.1 is a "local loopback" device. Processes can use this address in exactly the same way they use other network addresses.

**Try it out**

Execute the command `/sbin/ifconfig`. This will list the properties and current status of network devices in your system. You should see at least two entries: one for the Ethernet interface, and the other for `lo`, the local loopback device.

This command should work on both your development host and your target with similar results. `ifconfig`, with appropriate arguments, is also the command that sets network interface properties.

### The Server Process

Fig. 10.1 illustrates the basic steps that the server process goes through to establish communication. We start by creating a socket and then `bind()` it to a name or destination address. For local sockets, the name is a file system entry, often in `/tmp` or `/usr/tmp`. For network sockets, it is a *service identifier* consisting of a "dotted quad" Internet address (as in 192.168.1.11, for example) and a protocol port number. Clients use this name to access the service.

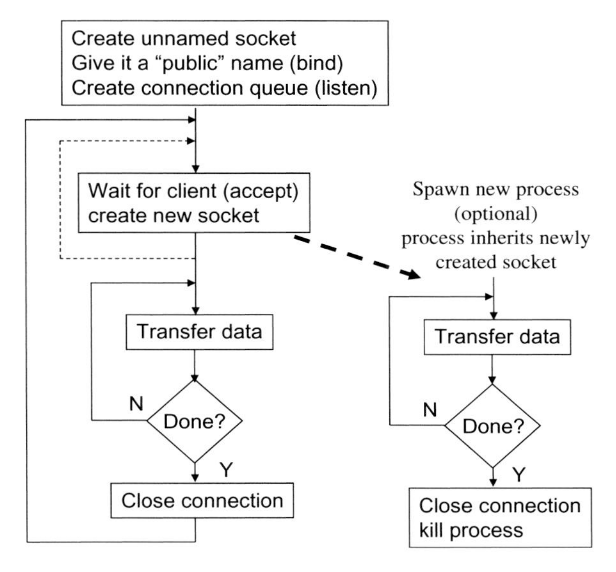


**Figure 10.1**
Network server.

Next, the server creates a connection queue with the `listen()` service, and then waits for client connection requests with the `accept()` service. When a connection request is received successfully, `accept()` returns a new socket, which is then used for this connection's data transfer. The server now transfers data using standard `read()` and `write()` calls that use the socket descriptor in the same manner as a file descriptor. When the transaction is complete, the newly-created socket is closed.

The server may very well spawn a new process or thread to service the connection while it goes back and waits for additional client requests. This allows a server to serve multiple clients simultaneously. Each client request spawns a new process/thread with its own socket. If you think about it, that is how a web server operates.

### The Client Process

Fig. 10.2 shows the client side of the transaction. The client begins by creating a socket and naming it to match the server's publicly advertised name. Next, it attempts to `connect()` to the server. If the connection request succeeds, the client proceeds to transfer data using `read`

() and write() calls with the socket descriptor. When the transaction is complete, the client closes the socket.

If the server spawned a new process to serve this client, that process should go away when the client closes the connection.

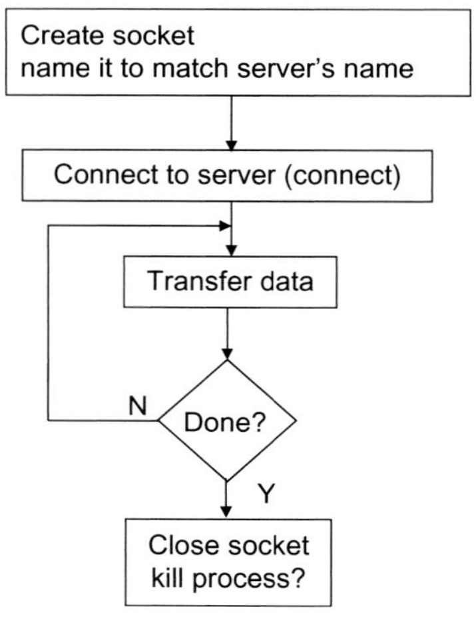


**Figure 10.2**
Network client.

### *Socket Attributes*

The socket system call creates a socket and returns a descriptor for later use in accessing the socket.

```
#include<sys/socket.h>
int socket (int domain, int type, int protocol);
```

A socket is characterized by three attributes that determine how communication takes place. The *domain* specifies the communication medium. The most commonly used domains are PF_UNIX for local file system sockets, and PF_INET for Internet connections. The "PF" here stands for Protocol Family.

The domain determines the format of the socket name or address. For PF_INET, the address is AF_INET, and is in the form of a dotted quad. Here "AF" stands for Address Family. Generally, there is a 1 to 1 correspondence between AF_ values and PF_ values. A network computer may support many different network services. A specific service is identified by a "port number." Established network services like ftp, http, etc., have defined port numbers below 1024. Local services may use port numbers above 1023.

Some domains, PF_INET for example, offer alternate communication mechanisms. SOCK_STREAM is a sequenced, reliable, connection-based, two-way byte stream. This is the basis for TCP, and is the default for PF_INET domain sockets. SOCK_DGRAM is a *datagram* service. It is used to send relatively small messages, with no guarantee that they will be delivered or that they won't be reordered by the network. This is the basis of UDP. SOCK_RAW allows a process to access the IP protocol directly. This can be useful for implementing new protocols directly in User Space.

The protocol is usually determined by the socket domain, and you do not have a choice. So, the protocol argument is usually zero.

## A Simple Example

### The Server

Time to create another Eclipse project. You know the drill. Call it "network." It is a makefile project, and this one is located in home/samples/src/network. Import the project settings with the include paths.

Open the file netserve.c in the editor. First we create a server_socket that uses streams. Next, we need to bind this socket to a specific network address. That requires filling in a sockaddr_in structure, server_addr. The function inet_aton() takes a string containing a network address as its first argument, converts it to a binary number, and stores it in the location specified by the second argument, in this case the appropriate field of server_addr. inet_aton() returns zero if it succeeds. In this example, the network address is passed in through the compile-time symbol SERVER so that we can build the server to run either locally through the loopback device, or across the network.

The port number is 16 bits, and is also passed in from the makefile through the compile-time symbol PORT. The function htons() is one of a small family of functions, macros actually, that solves the problem of transferring binary data between computer architectures with different byte ordering policies. The Internet has established a standard "network byte order," which happens to be Big Endian. All binary data is expected to be in network byte order when it reaches the network. htons() translates a short (16 bit) integer from "host byte order," whatever that happens to be, to network byte order. There is a companion function, ntohs() that translates back from network byte order to host order. Then there is a corresponding pair of functions that do the same translations on long (32 bit) integers.[2]

Now we *bind* server_socket to server_addr with the bind() function. Finally, we create a queue for incoming connection requests with the listen() function. A queue length of one

[2] Try to guess the names of the long functions.

should be sufficient in this case, because there is only one client that will be connecting to this server.

Now we are ready to *accept* connection requests. The arguments to `accept()` are:

- The socket descriptor.
- A pointer to a `sockaddr` structure that `accept()` will fill in.
- A pointer to an integer that currently holds the length of the structure in argument 2. `accept()` will modify this if the length of the client's address structure is shorter.

`accept()` blocks until a connection request arrives. The return value is a socket descriptor to be used for data transfers to/from this client. In this example, the server simply echoes back text strings received from the client until the incoming string begins with "q."

### The Client

Now look at `netclient.c`. `netclient` determines at run time whether it is connecting to a server locally or across the network. We start by creating a socket and an address structure in the same manner as in the server. Then we *connect* to the server by calling `connect()`. The arguments are:

- The socket descriptor.
- A pointer to the `sockaddr` structure containing the address of the server we want to connect to.
- The length of the `sockaddr` structure.

When `connect()` returns, we are ready to transfer data. The client prompts for a text string, writes this string to the socket, and waits to read a response. The process terminates when the first character of the input string is "q."

### Try it Out

We will need to create make targets for the server and client. Take a look at the makefile to see how these targets are specified. This project includes a number of make targets. You can create them all now, or wait until they are needed. Note, incidentally, that there is no all target "at all." Make both the client and the local server. Set up the host debug configuration to debug the server, and start a debug session. Step through the code up to and through the call to `accept()`, which will block until the client initiates a connection.

In a shell window, `cd` to the `network/` directory and execute `./netclient`. Type in a few strings and watch what happens. To terminate both processes, enter a string that begins with "q" ("quit" for example).

Next, we will want to run `netserve` on the target with `netclient` running on the host. Delete `netserve.o`, and build the make target for the remote server.

In the terminal window connected to the target (the one running `ssh`), `cd` to the `network/` directory and execute `./netserve`. Back in the original window, execute `./netclient remote`.

## A Remote Thermostat

Moving on to a more practical example, our thermostat may very well end up in a distributed industrial environment where the current temperature must be reported to a remote monitoring station, and setpoint and limit need to be remotely settable. Naturally we will want to do that over a network.

Copy `monitor.c` and `thermostat.c` from the `Posix/` directory to `network/`, and do a Refresh on the network project to bring those files into Eclipse. Actually, `thermostat.c` does not require any changes. Here are the modifications required in `monitor.c`:

1. Isolate the process of setting up the net server and establishing a connection in a separate function called `createServer()`, which consists roughly of lines 18 to 55 from `netserve.c`. The reason for doing it this way will become apparent soon. `createServer()` takes no arguments, and returns an `int` that is the socket number returned by `accept()`. It should return a −1 if anything fails.
2. Call `createServer()` from `monitor()` just before the `while (1)` loop, and check the return value for error.
3. Replace `fgets()` with `read()` on the socket returned by `createServer()`.
4. Add a new command to the `switch` statement, "?," that returns the current temperature over the network. Actually, we might want to query any of the thermostat's parameters. Think about how you might extend the command protocol to accomplish that.
5. The `netclient` expects a response string for every string it sends to the server. So return "OK" for every valid parameter change command, and return "???" for an invalid command.

Create make targets for both the simulation and target versions of the thermostat. Build the simulation version, and point the host debug configuration at `network/thermostat_s`. You will need two shell windows to exercise the thermostat: one for the `devices` program and the other for `netclient`.

When you are satisfied with the simulation, build the target version, and try it there. Remember to delete `thermostat.o` and `monitor.o`[3] before rebuilding. Also remember to execute `netclient remote`, to get the client talking to the server on the target.

[3] There should be a way to do this automatically in the makefile, but I haven't figured it out yet.

## Multiple Monitor Threads

As the net thermostat is currently built, only one client can be connected to it at a time. It is quite likely that in a real application, multiple clients might want to connect to the thermostat simultaneously. Remember from our discussion of sockets that the server process may choose to spawn a new process or thread in response to a connection request so it can immediately go back and listen for more requests.

In our case, we can create a new monitor thread. That is the idea behind the `createServer()` function. It can be turned into a server thread whose job is to spawn a new monitor thread when a connection request comes along. So give it a try.

Here, in broad terms, are the steps required to convert the net thermostat to service multiple clients:

1. Copy `monitor.c` to `multimon.c` and work in that file.
2. Recast `createServer()` as a thread function.
3. In the `createThread()` function, change the arguments of the `pthread_create()` call to create the server thread instead of the monitor thread.
4. Go down to the last four lines in `createServer()`, following the comment "Accept a connection." Bracket these lines with an infinite while loop.
5. Following the `printf()` statement, add a call to `pthread_create()` to create a monitor thread.
6. Add a `case "q":` to the `switch` statement in the monitor thread. "q" says the client is terminating (quitting) the session. This means the monitor thread should exit.

Now, here are the tricky parts. The `client_socket` will have to be passed to the newly-created monitor thread. Where does the thread object come from for each invocation of `pthread_create()`? A simple way might be to estimate the maximum number of clients that would ever want to access the thermostat, and then create an array of `pthread_t`s that big. The other, more general, approach is to use dynamic memory allocation, the `malloc()` function.

But the biggest problem is how do you recover a `pthread_t` object when a monitor thread terminates? The monitor thread itself cannot return it, because the object is still in scope, that is, still being used. If we ignore the problem, the thermostat will eventually become unresponsive either because all elements of the `pthread_t` array have been used, or we run out of memory because nothing is being returned to the dynamic memory pool.

Here are some thoughts. Associate a flag with each `pthread_t` object, whether it is allocated on a fixed array or dynamically allocated. The flag, which is part of a data structure that includes the `pthread_t`, indicates whether its associated `pthread_t` is free, in use, or "pending," that is, able to be freed. The server thread sets this flag to the "in use" state when creating the monitor. The structure should probably also include the socket number. It is this "meta" pthread_t object that gets allocated and passed to the monitor thread.

Now the question is, when can the flag be set to the "free" state? Again, the monitor cannot do it, because the `pthread_t` object is still in use until the thread actually exits. Here is where the third state, PENDING, comes in.

The monitor sets the flag to PENDING just before it terminates. Then we create a separate thread, call it "resource" if you will, that does the following:

1. Wakes up periodically and checks for monitor threads that have set their flags to PENDING
2. Joins the thread
3. Marks the meta pthread object free
4. Goes back to sleep

Remember that when one thread joins another, the "joiner" (in this case resource) is blocked until the "joinee" (monitor) terminates. So, when resource continues from the `pthread_join()` call, the just-terminated monitor's pthread_t object is guaranteed to be free. The semantics of `pthread_join()` are:

```
int pthread_join (pthread_t thread, void **ret_value)
```

It is quite likely that by the time the resource thread wakes up, the monitor thread has already exited, in which case `pthread_join()` returns immediately with the error code `ESRCH`, meaning that *thread* does not exist.

The `createThread()` function now needs to create the server thread and the resource thread. Likewise, `terminateThread()` needs to cancel and join both of those. And of course, any monitor threads that are running also need to be cancelled. That is, when the resource thread terminates, it needs to go through the list of meta pthreads checking for running monitors. Ah, but the resource thread doesn't know it is being canceled.

Here is a good use for a cleanup handler. Write a cleanup handler that runs through the list of meta pthreads looking for any that are in the IN USE state. Cancel and join the corresponding thread. Call `pthread_cleanup_push()` just before the main loop in the resource thread. Add a call to `pthread_cleanup_pop()` just after the loop. It will never be called, but it closes the block opened by `pthread_cleanup_push()`.

If you do not feel up to writing this yourself, you may have already noticed that there is an implementation of `multimon.c` in the `samples/.working/` directory.

There is a separate makefile called `Makefile.multi` to build the multiple client versions of `netthermo`. The build command for the Eclipse target is:

```
make -f Makefile.multi
```

Add `SERVER=REMOTE` to build the target version.

So how do you test multiple monitor threads? Simple. Just create multiple shell windows, and start `netclient` in each one. When you run thermostat under the debugger, you will note that a new thread is created each time `netclient` is started.

## Embedded Web Servers

While programming directly at the sockets level is a good introduction to networking, and a good background to have in any case, most real-world network communication is done using higher level application protocols. HTTP, the protocol of the World Wide Web, has become the de facto standard for transferring information across networks. No big surprise there. After all, virtually every desktop computer in the world has a web browser. Information rendered in HTTP is accessible to any of these computers with no additional software.

### Background on HTTP

HTTP is a fairly simple synchronous request/response ASCII protocol over TCP/IP, as illustrated in Fig. 10.3. The client sends a request message consisting of a header and, possibly, a body separated by a blank line. The header includes what the client wants, along with some optional information about its capabilities. Each of the protocol elements shown in Fig. 10.3 is a line of text terminated by CR/LF. The single blank line at the end of the header tells the server to proceed.

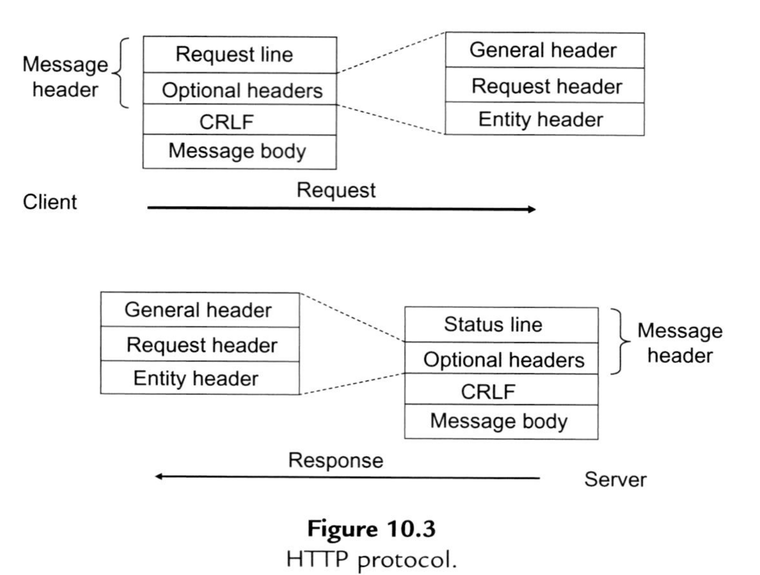


**Figure 10.3**
HTTP protocol.

A typical request packet is shown in Listing 10.1. The first line starts with a *method token*, in this case `GET`, telling the server what "method" to use in fulfilling this request. This is followed by the "resource" that the method acts on, in this case a file name. The server replaces the "/" with the default file `index.html`. The remainder of the first line says the client supports HTTP version 1.1.

```
GET / HTTP/1.1
Host: 127.0.0.1
User-Agent: Mozilla/5.0 (X11; U; Linux i586; en-US; rv:1.2.1) Gecko/20030225
Accept: text/xml,application/xml,application/xhtml+xml,text/html;q=0.9,text/plain;q=0.8
Accept-Language: en-us, en;q=0.50
Accept-Encoding: gzip, deflate, compress;q=0.9
Accept-Charset: ISO-8859-1, utf-8;q=0.66, *;q=0.66
Keep-Alive: 300
Connection: keep-alive
<blank line>
```

**Listing 10.1**

The `Host:` header specifies to whom the request is directed, while the `User-Agent:` header identifies who the request is from. Next come several headers specifying what sorts of things this client understands in terms of media types, language, encoding, and character sets. The `Accept:` header line is actually much longer than shown here.

The Keep-Alive: and Connection: headers are artifacts of HTTP version 1.0, and specify whether the connection is "persistent," or is closed after a single request/response interaction. In version 1.1 persistent connections are the default. This example is just a small subset of the headers and parameters available. For our fairly simple embedded server, we can in fact ignore most of them.

### *A Web-Enabled Thermostat*

To make data available via HTTP, you need a web server. Creating a web server for an embedded device is not nearly as hard as you might think. That is because all the rendering, the hard part, is done by the client, the web browser. By and large, all the server has to do is deliver files.

The `network/` directory contains a simple example called `webserve.c`. Open it in Eclipse. It starts out quite similar to `netserve.c`, except that it listens on port 80, the one assigned to HTTP. Once the connection is established, the server reads a request message from the client and acts on it. For our rather limited purposes, we are only going to handle two methods: POST and GET.

In the case of GET, the function `doGETmethod()` near line 183 opens the specified file, and determines its content type. In "real" web servers like Apache, HTML files are kept in a specific directory. In this case it just seems easier to put the files in the same directory as the program, so `doGETmethod()` strips off the leading "/" if present, to make the path relative to the current directory.

If everything is OK, we call `responseHeader()` to send the success response. The response header indicates the content type of the file, and also tells the server that we want to close the connection, even if the client asked to keep the connection alive. Finally, we send the file itself. If it is an HTML file, we need to parse it looking for dynamic content tags.

### *Dynamic Web Content*

Just serving up static HTML web pages is not particularly interesting, or even useful, in embedded applications. Usually, the device needs to report some information and we may want to exercise some degree of control over it. There are a number of ways to incorporate dynamic content into HTML, but, since this is not a book on web programming, we are going to take a "quick and dirty" approach.

This is a good time to do a quick review of HTML. Take a look at the file `index.html` in the `network/` directory.

A nice feature of HTML is that it is easily extensible. You can invent your own tags. Of course, any tag the client browser does not understand it will simply ignore. So if we invent a tag, it has to be interpreted by the server before sending the file out. We'll invent a tag called `<DATA>` that looks like this:

```
<DATA data_function>
```

`index.html` has one data tag in it.

The server scans the HTML text looking for a `<DATA>` tag. `data_function` is a function that returns a string. The server replaces the `<DATA>` tag with the returned string. Open `webvars.c`. Near the top is a table with two entries each consisting of a text string and a function name. Just below that is the function `cur_temp()`, which returns the current temperature of our thermostat as an ASCII string.

Now go back to `webserve.c`, and find the function `parseHTML()` around line 124. It scans the input file for the `<DATA>` tag. When one is found, it writes everything up to the tag out to the socket. Then it calls `web_var()` with the name of the associated data function. `web_var()`, in `webvars.c`, looks up and invokes the data function, and returns its string value. The return value from `web_var()` is then written out and the scan continues.

Needless to say, a data function can be anything we want it to be. This particular example happens to return the current temperature. Incidentally, there is a subtle bug in `parseHTML()`. See what you can do about it.[4]

### Forms and the POST Method

The <DATA> tag is how we send dynamic data from the server to the client. HTML includes a simple mechanism for sending data from the client to the server. You have no doubt used it many times while web surfing. It is the <FORM> tag. The one in our sample `index.html` file looks like this:

```
<FORM ACTION="control" METHOD="POST">
        <INPUT TYPE=TEXT NAME="Setpoint" SIZE=10>Setpoint:
        <INPUT TYPE=TEXT NAME="Limit" SIZE=10>Limit:
        <INPUT TYPE=TEXT NAME="Deadband" SIZE=10>Deadband:
        <INPUT TYPE="submit" NAME="Go">
</FORM>
```

This tells the browser to use the POST method to execute an ACTION named "control" to send three text variables when the user presses a "Go" button. Normally, the ACTION is a CGI script. That is beyond the scope of this discussion, so we'll just implement a function for "control."

Have a look at the `doPOSTmethod()` function around line 165. It retrieves the next token in the header, which is the name of the ACTION function. Then it calls `web_var()`, the same function we used for the dynamic <DATA> tag. In this case, we need to pass in another parameter, a pointer to the message body because that is where the data strings are.

In the <DATA> tag case, a successful return from `web_var()` yields a string pointer. In the POST method case, there is no need for a string pointer, but `web_var()` still needs to return a nonzero value to indicate success.

The body of an HTTP POST message contains the values returned by the client browser in the form:

```
<name1>=<value1>&<name2>=<value2>&<name3>=<value3>&Go=
```

The function `parseVariable()` in `webvars.c` searches the message body for a variable name, and returns the corresponding value string.

---

[4] Hint: What happens if a <DATA> tag spills over a buffer boundary? An expedient way around this is to read and write one character at a time. But that would incur a significant performance hit, because each call to `read()` and `write()` causes a transition to kernel space with the attendant overhead. So it is better if we can read and write in larger chunks.

### Build and Try it

Create the necessary make targets for the web thermostat, and try it out. Run it under the debugger on your workstation so you can watch what happens as a web browser accesses it. You must be running as root in order to bind to the HTTP port, 80.

In your favorite web browser enter:

http://127.0.0.1

into the destination window. You should see something like Fig. 10.4 once the file `index.html` is fully processed.

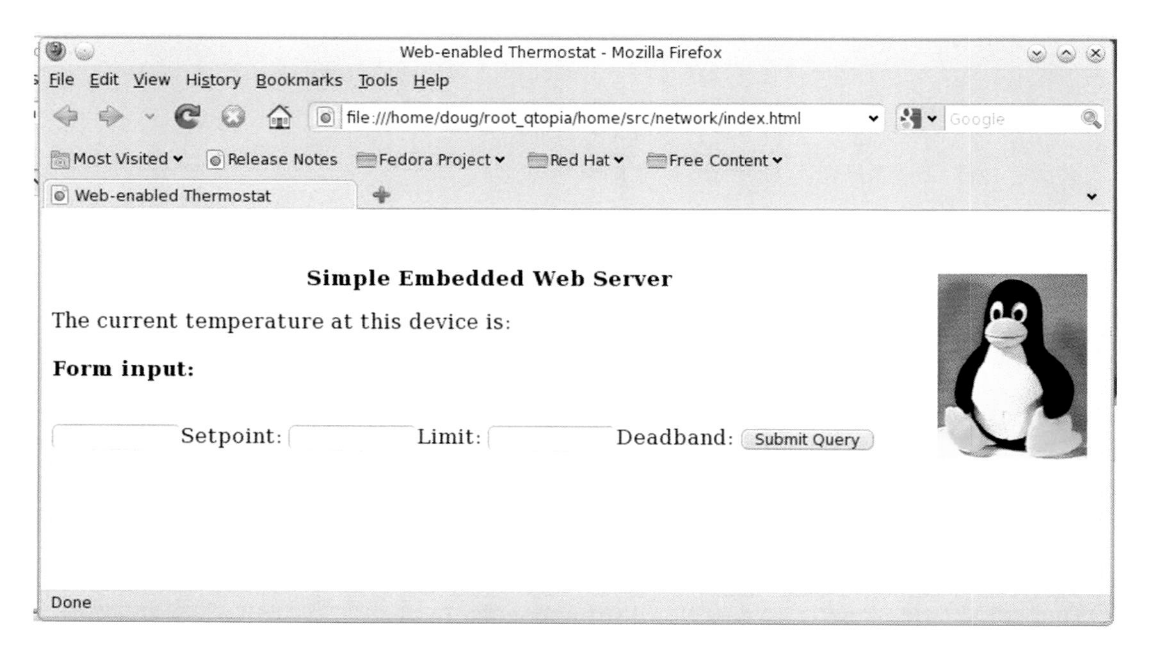


**Figure 10.4**
Web thermostat.

When running the web thermostat on the target board, enter the IP address of your BBB into your web browser.

Once you have that running on both the workstation and target, here is an "extra credit" assignment. There are three small GIF files in the `network/` directory that are circles—one red, one green, and one open. These can be used to represent the states of the cooler and alarm. So try adding a couple of <DATA> tags that will put out the appropriate circle image based on the cooler and alarm states.

Of course, the other thing we would want to do is dynamically update the temperature reading. Dynamic web content gets into java applets, CGI, and other issues that are beyond the scope of this book, but are definitely worth pursuing. Google the phrase "embedded web server" to find a veritable plethora of commercial products and open source projects that expand on what we have done here.

## "Real" Web Servers

While `webserve.c` is a useful introduction to HTTP processing, it is rather limited. Fortunately, the open source world includes a number of practical web servers that are suitable for embedded applications. The Resources section includes a link to a Wikipedia article that provides a very useful comparison of web servers. Many of these are specifically intended for embedded use, while others were developed as a proof-of-concept to see just how small one could make a web server.

### Apache

Apache is a full-featured, open source, web server that, as of July 2016, runs 46% of the world's active websites, and 43% of the top million websites.[5] It also happens to be installed in our BBB root file system and running. If you point your web browser at your target board's IP address, you will see a web page that talks about BoneScript.

### Lighttpd

`lighttpd` seems to be a popular lightweight web server in the BeagleBone community. It is definitely small. The tar file is a mere 645 KB. `lighttpd` follows the usual build pattern:

```
./configure
make
su
make install
```

The executables, `lighttpd` and `lighttpd-angel`, end up in your workstation's `/usr/local/sbin`, and the libraries in `/usr/local/lib`. So move everything into the same locations in your target file system. It is possible I overlooked some configuration variable that will put everything in the right place to begin with.

Unfortunately, when I ran `lighttpd` I encountered a problem with glibc versions.

[5] https://en.wikipedia.org/wiki/Apache_HTTP_Server

*Node.js*

Node.js is not itself a web server, but it is claimed that you can use it to easily build web servers. Node.js is described as a server-side platform built on Google Chrome's JavaScript Engine (hence the js). It is intended for easily building fast and scalable network applications written in JavaScript and using an event-driven, nonblocking I/O model.

Node.js has been around since 2009, and is apparently fairly popular in the world of web programming. The BBB includes Node.js. However, since I know virtually nothing about JavaScript, I'm not going to try to describe it. The web page served up by the BBB describes a JavaScript library called BoneScript that, among other things, provides another way to access the hardware. The demos are probably a good place to start.

We have pretty much wrapped up our tour of application-level programming. In the next chapter, we dive down into Kernel space, and have a look at the Linux kernel itself.

## Resources

*Linux Network Administrators' Guide*, available from the Linux Documentation Project, www.tldp.org. Not just for administrators, this is a quite complete and quite readable tutorial on a wide range of networking issues.

Jones, M. Tim, *TCP/IP Application Layer Protocols for Embedded Systems*, Charles River Media, 2002. The idea of adding a <DATA> tag to HTML came from this book. It covers a wide range of application-level network protocols that can be useful in embedded scenarios.

Comer, Douglas, *Internetworking with TCP/IP, Vols. 1, 2,* and *3*, Prentice-Hall. This is the classic reference on TCP/IP. Volume 1 is currently up to its fifth edition dated 2005. Volume 2, coauthored with David L. Stevens, is at the third edition, 1998, and volume 3, also coauthored with Stevens, dates from 1997. Highly recommended if you want to understand the inner workings of the Internet.

Internet protocols are embodied in a set of documents known as "RFCs," Request for Comment. The RFCs are now maintained by a small group called the RFC Editor. The entire collection of RFCs, spanning the 30-plus year history of the Internet, is available from www.rfc-editor.org. In particular, HTTP is described in RFC 2616, SMTP is described in RFC 821, and POP3 is described in RFC 1081.

en.wikipedia.org/wiki/Comparison_of_web_server_software – This article features a pair of quite useful tables comparing more than a dozen small web server implementations.

httpd.apache.org – Home page for the Apache web server.

www.lighttpd.net – Home page for lighttpd.

nodejs.org/en/ – Home page for node.js.

CHAPTER 11

# Graphics programming with QT

**Chapter Outline**



*We are using Linux daily to UP our productivity – so UP yours!*
***Adapted from Pat Paulsen by Joe Sloan***

A great many embedded devices today have GUIs, graphical user interfaces, some of them quite sophisticated. In this chapter we will explore one approach to developing graphical interfaces, the QT library, and associated tools.

QT (pronounced "cute," not "cutee") is a cross-platform application framework used for developing applications that can run on a variety of platforms with little or no change to the underlying code base. While it is primarily aimed at developing graphical applications, it is also useful for developing command line tools and server consoles. QT is available in both an open source version licensed under LGPL 2.1, LGPL 3.0, and GPL 3.0, and a commercial version from the QT Company of Finland.

## Getting and Installing QT

The open source QT library is available from the CentOS 7 repositories. As root user, execute the following:

```
yum install qt
yum install qt-devel
```


Linux for Embedded and Real-time Applications.
DOI: http://dx.doi.org/10.1016/B978-0-12-811277-9.00011-0

`qt` installs a large number of libraries in `/usr/lib64` plus a subfolder, `qt4/`, under `/usr/lib64`. Among other things, `qt-devel` installs several executables in `/usr/bin`:

```
designer-qt4
lrelease-qt4
moc-qt4
uic-qt4
linguist-qt4
lupdate-qt4
qmake-qt4
```

Add `/usr/lib64/qt4/bin` to the front of your `PATH` in `.bash_profile`. Log out and log back in to make the new path effective.

The version available from CentOS is 4.8.5. The latest version as of June 2017 is 5.9. If you would like to play with that one, go to the download page listed in the resources section.

## *QT Basics*

QT is a collection of C++ classes that implement a wide range of graphical *widgets*. QT starts with the standard C++ object model, and then adds several significant features that support GUI programming:

- A mechanism for seamless object communication called *signals* and *slots*.
- Queryable and designable object properties.
- Events and event filters.
- Contextual string translation for internationalization.
- Interval driven timers that integrate many tasks in an event-driven GUI.
- Hierarchical and queryable object trees that organize object ownership.
- Guarded pointers (QPointer) that are automatically set to 0 when the referenced object is destroyed, unlike normal C++ pointers that become dangling pointers when their objects are destroyed.
- A dynamic cast that works across library boundaries.

Many of these features are implemented with standard C++ techniques, based on inheritance from `QObject`. Others, like the object communication mechanism and the dynamic property system, require the Meta-Object System provided by QT's own Meta-Object Compiler (moc). Collectively, these features are known as the QT Object Model.

### Signals and Slots

In GUI programming, when one widget changes, say by clicking on a button, we often want another widget to be notified. If a user clicks a **Close** button, we probably want the window's `close()` function to be called.

Older toolkits achieve this kind of communication using *callbacks*. A callback is a function that will be invoked when a processing function needs to notify you that some event has occurred. This is typically done by passing a pointer to the callback to the processing function.

QT's alternative to callbacks is signals and slots. A *signal* is emitted when a particular event occurs. QT's widgets have many predefined signals, but you can always subclass widgets to add your own signals to them. A *slot* is a function that is called in response to a particular signal. Widgets have many predefined slots, but it is common practice to subclass widgets and add your own slots, so that you can handle the signals that you are interested in.

Signals and slots are loosely coupled: a class that emits a signal neither knows nor cares which slots, if any, receive the signal. QT's signals and slots mechanism ensures that if you connect a signal to a slot, the slot will be called with the signal's parameters at the right time. Signals and slots can take any number of arguments of any type. They are completely type safe.

All classes that inherit from `QObject` or one of its subclasses, `QWidget` for example, can contain signals and slots. An object emits a signal when its state changes in a way that may be interesting to other objects. This is all the object does to communicate. It does not know or care whether anything is receiving the signals it emits. This is true information encapsulation, and ensures that the object can be used as a software component.

Slots can be used for receiving signals, but they are also normal member functions. Just as an object does not know if anything receives its signals, a slot does not know if it has any signals connected to it. This ensures that truly independent components can be created with QT.

You can connect as many signals as you want to a single slot, and a signal can be connected to as many slots as you need. It is even possible to connect a signal directly to another signal. This will emit the second signal immediately, whenever the first is emitted. Fig. 11.1 illustrates this.

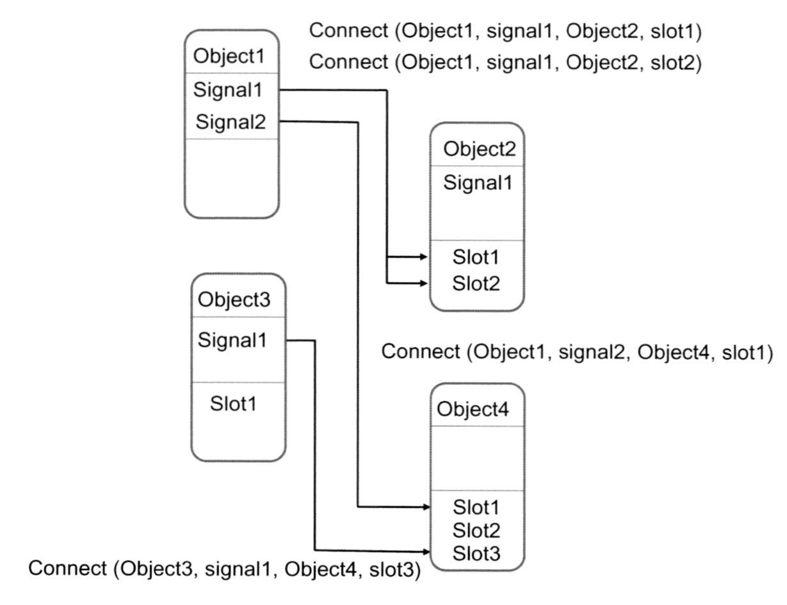


**Figure 11.1**
Signals and slots.

## *A Simple Example*

To get a feel for how QT works, let us start with a simple example of a text edit window. The source code is shown in Listing 11.1. Each QT class has its own header file with the same name as the class. The header `QtGui` in line 2 pulls in all of the GUI class headers.

Line 6 creates a `QApplication` object. This object manages application-wide resources, and is necessary to run any QT program that has a GUI. It needs `argv` and `args`, because QT accepts a few command line arguments. Line 8 creates a text edit widget, and line 9 creates a push button widget with the text "Quit." Then, in line 11 we connect the `clicked()` signal of the `quitButton` to the `quit()` slot of the application. The `quit()` function terminates the program.

Line 13 creates a Vertical Box Layout widget. Widgets can contain other widgets. In this case, lines 14 and 15 make `textEdit` and `quitButton` children of `layout`. The layout widget organizes its children in a window, in this case vertically. Line 18 puts `layout` in a window, and line 20 displays the window. Finally, line 22 causes the `QApplication` to enter its event loop. QT applications are event-driven. Events are generated and sent to the application's widgets. Examples of events are mouse presses and key strokes. When you type text in the text edit widget, it receives key pressed events, and responds by drawing the text typed.

```
1
2   #include <QtGui>
3
4   int main (int argv, char **args)
5   {
6       QApplication app (argv, args);
7
8       QTextEdit *textEdit = new QTextEdit;
9       QPushButton *quitButton = new QPushButton ("&Quit");
10
11      QObject::connect (quitButton, SIGNAL(clicked()), qApp, SLOT(quit()));
12
13      QVBoxLayout *layout = new QVBoxLay out;
14      layout->addWidget (textEdit);
15      layout->addWidget (quitButton);
16
17      QWidget window;
18      window.setLayout (layout);
19
20      window.show();
21
22       return app.exec();
23  }
```

**Listing 11.1**
Text edit program.

Make a subfolder under your home directory, and create the program shown in Listing 11.1. To build the program, run the following commands:

```
qmake –project
qmake
make
```

The first command creates a *project* file with the extension `.pro`. This contains some elementary information about the project. The second command creates a rather elaborate makefile, and the last command builds the target.

## *QT Designer*

Yes, you can build elaborate GUIs by manually programming as we just did, but there is a better way. QT has a graphical design tool called *Qt Designer* that allows you to design in a what-you-see-is-what-you-get environment, and generates the code for you.

There are a couple of ways to start Qt Designer. In a command shell, enter `designer-qt4`. In CentOS 7 it is also available from the Application Launcher under `Development>Qt4`

`Designer`. The initial dialog offers four options: open a form, open a recently edited form, cReate a form, or Close. At this point, you will want to create a form.

That brings up the main designer window, as shown in Fig. 11.2. At the top are the usual menu and tool bars. On the left side is the *widget* box, with a list of all the widgets you can put in your GUI. In the center is the main graphical editing window. On the right are three windows: the *Object Inspector* is a tree structure showing the relationship of all the objects in your GUI. Below that is the *Property Editor*. Select an object in the Object Inspector, and the properties of that object show up in the Property Editor where you can edit them. Below that is a window with three tabs: *Signal/Slot Editor*, *Action Editor*, and *Resource Browser*. We will deal with them later.

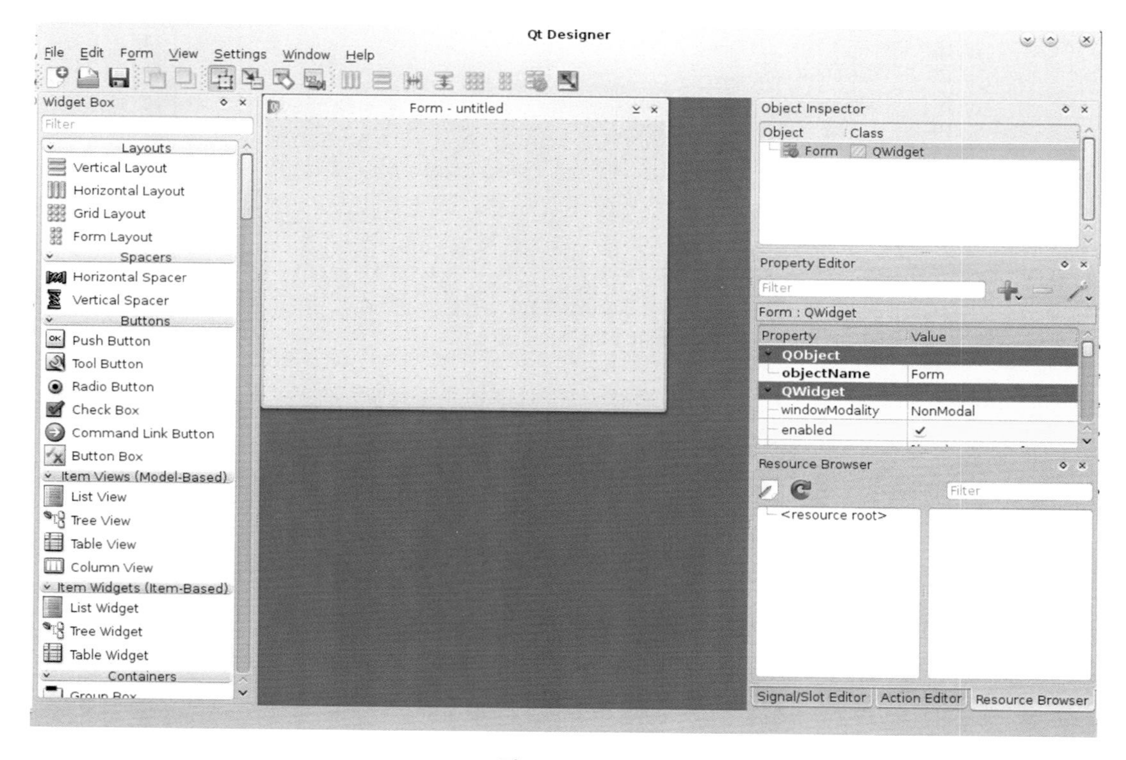


**Figure 11.2**
Qt Designer.

For now, we simply want to recreate the text edit example we did earlier to get a feel for how Qt Designer works, and what it creates. Right now there is one object in our form, a `QWidget`, which is always the top-level class on a form. Currently the `objectName` is "Form," not very descriptive. In the Property Editor, change it to something sensible like textEdit. Then scroll down to find `windowTitle`, and change it to something like "Edit Example" maybe.

To add widgets to the form, just drag them from the Widget Box to the form. Scroll down to find the Text Edit widget, and drag it to the form. Scroll back up to find the Push Button widget, and drag one of those to the form. Notice that the Object Inspector shows the two new widgets as children of QWidget.

Select the Push Button in the Object Inspector, and change the `objectName` to "quitButton." Now scroll down in the Property Editor to find the `text` property. Change that to "Quit."

We are going to add one more feature to this example: a second button that will clear the text edit window. This will require us to provide the slot function to do the clearing. Drag another button on to the form, change the `objectName` to "clearButton," and change the text to "Clear."

Next we want to put our widgets into a vertical layout. With the Clear button selected, hold the Ctrl key and select the other two widgets. Now all three widgets are selected. In the `Form` menu, select `Lay Out Vertically`. Note that the widgets are now children of `QVBoxLayout`.

Now we need to connect a signal of the Quit button to a slot of the Text Edit window. In the `Edit` menu, select `Edit Signals/Slots`. Select the Quit button, and drag toward the Text Edit window. That brings up the dialog shown in Fig. 11.3. Check `Show signals and slots inherited from QWidget`, and select `clicked()`. The slot window now becomes active. Oddly, `quit()` is not one of the options. Select `close()` instead, and click `OK`. To see what the final form looks like, select `Form>Preview...` (Fig. 11.4).

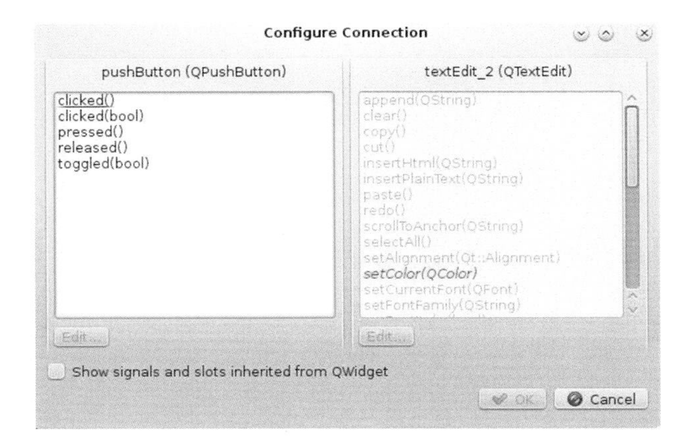


**Figure 11.3**
Signals/Slots dialog.

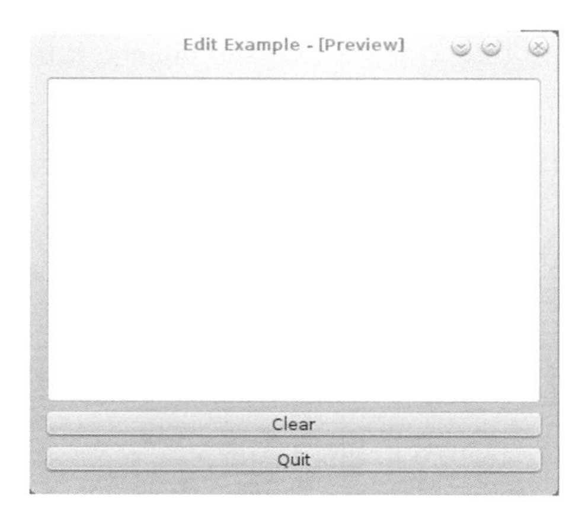


**Figure 11.4**
UI form preview.

Save the new form in a new subdirectory. It is saved in a .ui file, meaning user interface. Open that file with a text editor, and you will see that it is XML. In order to use this UI in a program, it must be translated into a C++ header file by QT's `uic`, user interface compiler. This step is handled by the makefile that is created by `qmake`.

Finally, we need a C++ source file that actually invokes the user interface. In this case, we need to create a new class that inherits from `QWidget` so we can declare a slot function to clear the text window in response to a click on the Clear button. Listing 11.2 shows the header file that prototypes our class.

```
1
2   #ifndef MYQTAPP
3   #define MYQTAPP
4
5   #include "ui_TextEdit.h"
6
7   class myQtApp : public QWidget, private Ui_editExample
8   {
9      Q_OBJECT
10
11   public:
12      myQtApp (QWidget *parent = 0);
13      ~myQtApp ();
14
15   public slots:
16      void clearText();
17  };
18   #endif
```

**Listing 11.2**
myQtApp.h.

The first thing the `myQtApp.h` does is include the user interface header at line 5. Then we declare the class in line 7. `editExample` is the name I gave to the top-level UI object. If you named yours something else, put that name here and prepend with `Ui_`. `Q_OBJECT` is a macro that expands to some definitions needed for signals and slots. At lines 12 and 13 we declare the constructor and destructor, and finally we declare the slot function that we will link to the Clear button.

Listing 11.3 shows the corresponding C++ file that implements our class and the main program. `setupUi()` in the constructor does exactly what its name implies. It sets up, but does not display, the user interface. At line 7 we connect the Clear button's `clicked()` signal to our slot function that will clear the screen. The destructor does not do anything, but it must be defined or you are likely to encounter a very obscure C++ linker error described below.

The slot function simply invokes the `clear()` member of the `textEdit` widget.

```
1
2  #include "myQtApp.h"
3
4  myQtApp::myQtApp (QWidget *parent)
5  {
6     setupUi(this); // this sets up GUI
7     connect (clearButton, SIGNAL(clicked()), this, SLOT(clearText()));
8  }
9  myQtApp::~myQtApp ()
10 {
11 }
12
13 void myQtApp::clearText ()
14 {
15    textEdit->clear();
16 }
17
18 int main (int argc, char *argv[])
19 {
20    QApplication app (argc, argv);
21    myQtApp *dialog = new myQtApp;
22
23    dialog->show();
24    return app.exec();
25 }
```

**Listing 11.3**
main.cpp.

`main()` starts out as our previous program did by creating the `QApplication` object to manage application resources. At line 21 we instantiate our class. Line 23 displays the UI, and line 24 enters the application's event loop.

The same three commands are used to build the program that we saw above, namely:

```
qmake -project
qmake
make
```

You usually do not need to rerun `qmake -project` unless you add or delete files in the project folder. On the other hand, it is a good idea to run both `qmake` and `make` whenever you edit any source files. The final executable is named for the project directory, not the C++ file.

Now about that obscure error I mentioned earlier. This is a fairly simple program, consisting as it does of just one source file. So it seemed to me to be quite sensible to declare and implement the myQtApp class directly in the source file, and eliminate the need for a header file. But doing so caused the linker to emit this error:

```
undefined reference to 'vtable for myQtApp'
```

This apparently happens a lot, because there are plenty of forum posts about it. The problem arises because one or more virtual methods that are not pure are not defined. I tried several approaches to fixing the problem, until I stumbled across a forum post that effectively said you must have a header file. So do not be surprised if you run across that error sometime.

## Putting QT on the ARM

Our journey up to now in this chapter has been to introduce the concepts of QT. This is easier done in the workstation environment. Now it is time to move our expertise to the BeagleBone target board. There are basically three steps involved:

1. Decide how to display graphics output from the BBB. You can get a display cape. This is pretty much your choice. There are a couple of options: capes where the display is built right on to the cape, and those where the display is connected via a cable to the cape. In the first case, you lose access to the pot that provides temperature to our thermostat example. The alternative is to use the VNC (Virtual Network Computing) protocol to display the output on your workstation.
2. Build and install the QT libraries on the BBB.
3. Modify the application build environment to build applications for the target.

### Build the Target Libraries

Start by getting the source code of the open source version of Qt. To get the source code, go to the QT download page listed in Resources, and click **Get your open source package**. Assuming you are using the 4.8.5 version of QT distributed with CentOS 7, scroll down that page to the line that says "Qt 5.8 and all older versions of Qt are available in the archive." Click the link in that line, and select the 4.8 directory. Then select the 4.8.5 directory. Download `qt-everywhere-opensource-src-4.8.5.tar.gz` to your home directory, or wherever you would like.

Untar `qt-everywhere-opensource-src-4.8.5.tar.gz`. Not surprisingly, that will create a subdirectory named `qt-everywhere-opensource-src-4.8.5`. The first step is to configure the package by executing the configure script. There are several required options and some "optional" options:

```
./configure -embedded arm-gnueabi -little-endian
```

Other options you might want to add include:

| | |
|---|---|
| `-qt-gfx-vnc.` | Build the VNC protocol driver |
| `-qt-gfx-fb` | If you have a display cape, this builds the framebuffer driver |
| `-qt-mouse-tslib` | If your display cape has a touchscreen, this enables the touchscreen library |
| `-prefix <path>` | Where to install QT. Default is /usr/local/Trolltech/QtEmbedded-4.8.5-arm |
| `-no-qt3support` | Unless you are supporting legacy applications, you probably do not need QT 3 support. |

The configure process takes a long time, much of it configuring examples and demos. If you do not need those, and perhaps don't need documentation either, you can add these options:

```
-nomake examples -nomake demos -nomake docs
```

You can get a complete list of options with this command:

```
./configure -embedded -help
```

You may want to put your configure command line into a shell script, as you may be building more than once. Now run your `configure` command. You will be asked whether you want to build the commercial or open source version of QT, and then asked to accept the terms of the open source licenses.

There is one more step before we can actually build QT. The prefix for the cross-tool chain names does not match what QT is expecting. Go to `qt-everywhere-opensource-src-4.8.5/mkspecs/qws/linux-arm-gnueabi-g++` and open the file `qmake.conf` with an editor. Note that there are seven lines with `arm-none-linux-gnueabi-`, and the name of a GCC tool. Change `none` to `unknown` and save the file.

Back in the `qt-everywhere-opensource-src-4.8/` directory, execute `gmake`. When that finishes, execute `qmake install`. That puts the final product in your chosen target directory. If the install destination is not in your target file system, then you need to copy the libraries to the target file system. Add an environment variable, `QTLIB`, to the BBB that points to the QT libraries.

If your display cape has a touchscreen, you will probably also want to get and build tslib, the touchscreen library. Get tslib from github with this command executed in your home directory:

```
git clone https://github.com/kergoth/tslib.git tslib
```

`cd` to the newly created `tslib/` and execute the following:

```
export CROSS_COMPILE=arm-unknown-linux-guneabi-
export CC=arm-unknown-linux-guneabi-gcc
export CXX=arm-unknown-linux-guneabi-g++
./autogen.sh
./configure –host=arm-linux-guneabi –prefix=~/tslib/tslib
make
make install
```

There is an error in the `make` step. If you have gotten this far in the book, you have probably already googled a few error messages. And if you have not, here is a good time to try it. The solution is relatively obscure, but pretty simple to implement. After `make install` completes, copy the `tslib/bin` and `tslib/lib` directories to `/usr` in your target file system. Copy `tslib/etc/ts.conf` to `/etc` in the target file system.

### *Create Build Environment for Target Applications*

To build QT applications for the BBB, you must use the tools in the version of QT that we built for the ARM. Change your `PATH` variable to point to where you installed QT, rather than `/uar/lib64/qt4/bin`. I used the default location, `/usr/local/Troltech/QtEmbedded-4.8.5-arm/bin`.

There are three variables that need to be added to the project file:

```
QMAKE_CC = arm-unknown-linux-gnueabi-gcc
QMAKE_CXX = arm-unknown-linux-gnueabi-g++
QMAKE_LINK = arm-unknown-linux-gnueabi-g++
```

Remember to add these variables each time you create a project file. There should be a way to do this automatically through `qmake`'s configuration, but I have not figured it out yet.

Even though you are now using tools from the ARM version of QT, Qt Designer is still the one that came with the CentOS version.

## A Graphical Thermostat

You can probably guess where all of this is leading. Fig. 11.5 is a QT form for our thermostat. It has five widgets: an LCD number widget for displaying temperature, two LED widgets to indicate high temperature and alarm, and two corresponding label widgets.

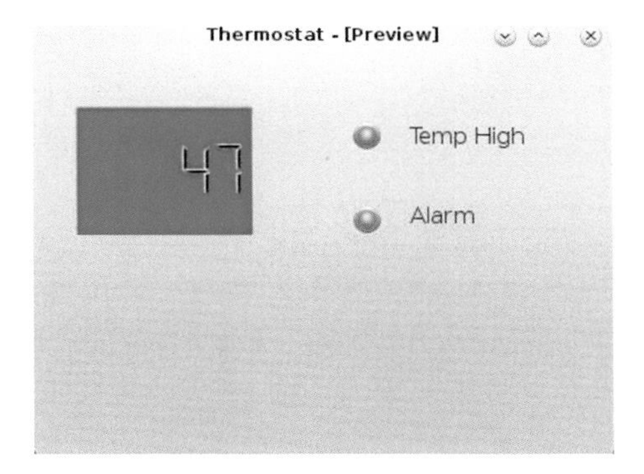


**Figure 11.5**
Graphical thermostat.

Your thermostat program will need to send a signal to the display slot of the LCD number widget. Interestingly, the LED widget, class KLed, is not documented in the QT documentation as it is really a part of KDE. Google "kled widget" to find the appropriate references.

After you get the basic display working, try adding some input widgets to manage setpoint, limit, and deadband.

This tour of graphical programming using QT concludes the second part of the book on embedded applications. In the next chapter, we will start an exploration of various components and tools to support embedded Linux development, starting with configuring and building the Linux kernel itself.

## Resources

qt.io – Home page for the QT company. There is a fair amount of documentation here, although I'm a little less than impressed with its organization. Of course YMMV.

info.qt.io/download-qt-for-application-development – This is the page for downloading the latest open source version of QT. From here you can also get to the QT archives where you can get the source code to version 4.8.5.

PART III

# *Components and tools*

CHAPTER 12

# Configuring and building the Linux kernel

## Chapter Outline



*Hackito ergo sum*

***Anonymous***

One of the neatest things about Linux is that you have the source code. You are free to do whatever you want with it. Most of us have no intention, or need, to dive in and directly hack the kernel sources. But access to the source code does mean that the kernel is highly configurable. That is, you can build a kernel that precisely matches the requirements, or limitations, of your target system.

Now again, if your role is writing applications for Linux, as we have been doing in the last few chapters, or if you are a Linux system administrator, you may never have to touch the kernel. But, as an embedded systems developer, you will most certainly have to build a new kernel, probably several times, either for the workstation or the target environment. Fortunately, the process of configuring and building a kernel is fairly straightforward.

Linux for Embedded and Real-time Applications.
DOI: http://dx.doi.org/10.1016/B978-0-12-811277-9.00012-2

My experience has been that building a new kernel is a great confidence building exercise, especially if you are new to Linux.

The remainder of this chapter details and explains the various steps required to configure and build an executable kernel image for our target board.

## Getting Started

### Where's the Source Code? Upstream versus Downstream Kernels

The mainline source code for the Linux kernel resides at www.kernel.org, where you can find the code for every version of the kernel back to Linus' original release in 1991. The sources are available in both gzipped tar format (.tgz), the slightly smaller bzipped format (.bz2), and the newer xz format (.xz).

`kernel.org` is where all kernel development takes place. Code stored here is referred to as the *upstream* kernel. Development at kernel.org is continuous, with a new version of the kernel appearing about every 6 weeks to 2 months. As of May 16, 2017, the latest stable release is 4.11.1 released on May 14.

It is perfectly OK to use an upstream kernel from kernel.org, provided it supports your hardware and the features of your environment. However, Linux distributors such as Red Hat and Ubuntu, as well as groups supporting embedded development such as BeagleBoard, take an upstream kernel and modify it for their particular purposes. The resulting kernel is referred to as a *downstream* kernel, which is generally distributed as a set of patches to a specific upstream kernel.

The process is illustrated in Fig. 12.1, where the vertical arrow represents the ongoing development on the upstream kernel at kernel.org. At some point, the BeagleBoard developers decided on version 3.8.13, made the necessary modifications, and released their patch set, which we will be using shortly. Likewise, the Red Hat folks settled on 3.10.0 for Red Hat Enterprise Linux 7 (CentOS 7), and made their modifications to that upstream kernel.

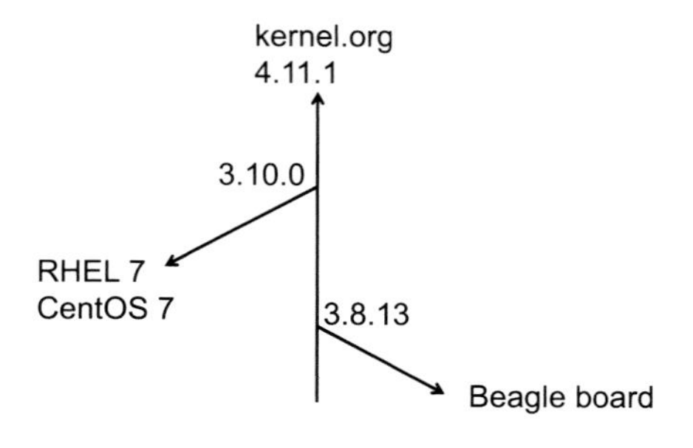


**Figure 12.1**
Upstream and downstream kernels.

### *Kernel Version Numbering*

Generally, Linux sources are installed as subdirectories of `/usr/src/kernels`. The subdirectories usually get names of the form:

```
linux-<version_number>-<additional_features>.
```

`<version_number>` identifies the base version of the kernel as obtained from kernel.org, and looks something like this: 3.8.13. The first number is the "version," in this case 3. Normally, this number increments only when truly major architectural changes are made to the kernel and its APIs. The version number was 2 from 1996 until mid-2011, when Linus arbitrarily rolled the version to 3 in honor of the 20th anniversary of Linux.

The second number, 8, is used to identify releases where the kernel APIs may differ, but the differences are not enough to justify a full version change. New releases appear about every 2 to 3 months.

The final number, 13 in this example, represents bug and security fixes that do not change the kernel APIs. Applications built to a specific release should, in principle, run on any security level.

`<additional_features>` is a way of identifying patches made to an upstream kernel to support additional functionality. The `<additional_features>` field is also used by the kernel developers to identify developmental source trees known as *release candidates*. Starting from a stable kernel release, sufficiently stable new code is merged into the mainline to become the next release `–rc1`. So, for example, `3.8.13` would become `3.9-rc1`. A period of 6 to 10 weeks of testing and patching ensues, with the `-rc` number being rolled about every week. When the new tree is deemed sufficiently stable, the `–rcn` is dropped and the process starts over again.

Given that you may have multiple kernel source trees in your file system, the kernel you are currently working with is usually identified by a symbolic link in `/usr/src/kernels` called `linux`.

### *Additional Tools*

Before we actually install the kernel source tree, there are some additional tools we need to install in case they are not already on your system. You will, of course, need to execute these steps as root user.

- `yum install qt-devel` — This is a graphical environment used by the kernel's `make xconfig` script.

- `yum install lzop` — A file compression program that the kernel scripts use.
- `yum install sudo` — This is another way to become root user, as was discussed in Chapter 3, Introducing Linux. Some of the BeagleBoard scripts make use of `sudo`.

You will need to edit the file `/etc/sudoers` (you are still root user right?) to allow your normal user to `sudo`. Good practice says you should only edit `/etc/sudoers` with `visudo`, a special version of `vi` that checks the syntax of the file before saving, because editing the file wrong could mess up your system. For what it is worth, I've edited `/etc/sudoers` with `kwrite` and gotten away with it.

Find the line that reads:

```
root ALL=(ALL) ALL
```

Copy that line and change `root` to your user name. Save the file.

### Getting and Installing the BeagleBone Kernel

Installing an upstream kernel involves little more than downloading and untarring the archive file for the desired version. Installing a downstream kernel is a little more complicated because, in addition to downloading the upstream source, we have to apply the appropriate patches.

We will use the source code control tool `git` to get and patch the kernel. We will go into `git` in more detail in a later chapter. For now, just follow these commands:

- Decide where you want to install the source tree. While the convention is `/usr/src/kernels`, there is nothing wrong with installing it under your home directory.
- `cd` to the chosen directory.
- `git clone git://github.com/beagleboard/kernel.git`
- `cd kernel`
- `git checkout 3.8`
- `./patch.sh`

This last step can take a *long* time depending on the speed of your processor, and the number of cores. When it is finished, you will have the directory structure shown in Fig. 12.2. The most important subdirectory for our purposes is `kernel/`, which holds the actual kernel source tree.

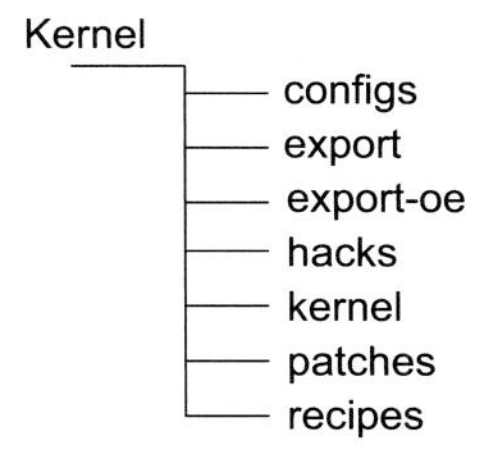


**Figure 12.2**
BeagleBone kernel structure.

Finally, we need to get a power management firmware file from another source:

- `wget http://arago-project.org/git/projects/?p=am33x-cm3.git\;a=blob_plain\;f=bin/am335x-pm-firmware.bin\;hb=HEAD -O kernel/firmware/am335x-pm-firmware.bin`

## *Patching the Kernel*

The mechanism for changing released source code in an orderly manner is the `patch` utility. The input to `patch` is a text file created by the `diff` utility that compares two files and reports any differences. So, when an Open Source programmer wants to distribute an upgrade to released source code, she does a `diff` between the modified code and the original code, redirecting the output of `diff` to a patch file. Something like this:

```
diff –uprN original_code new_code> file.patch
```

The `.patch` extension is not required, but is the convention. See the `diff man` page to learn what the various options represent.

Then, to apply the patch, copy the patch file to the directory containing the original code and execute:

```
patch –p1< file.patch
```

Note that `patch` normally takes its input from `stdin`, so we have to redirect input to the file. The `–p1` flag tells `patch` to remove one slash and everything before it from the names of files to be patched. This recognizes the usual situation that the root directory from which the patch was created is not the same as the directory in which we are applying it. Removing everything in the path before the first slash makes the directories relative.

You could have `diff` create one gigantic patch file for the entire kernel source tree. The BeagleBone developers chose not to do that, opting instead to create individual patches for each file that is modified. This adds up to over a 1000 files in the `patches/` subdirectory. The motivation for this is that it is probably easier to distribute updates to individual files.

To get a better feel for patch files, have a look at one or more of the files under the `patches/` subdirectory. Try, for example, `arm/0001-deb-pkg-Simplify-architecture-matching-for-cross-bui.patch`. This happens to be a patch to a shell script.

## The Kernel Source Tree

Needless to say, the kernel encompasses a very large number of files—C sources, headers, makefiles, scripts, etc. The BeagleBone version 3.8.13 kernel has 48,683 files, taking up 1.7 GB. So, not surprisingly, there is a standard directory structure to organize these files in a manageable fashion. Fig. 12.3 shows the kernel source tree starting at `kernel/kernel/`. The directories are as follows:

`.git` – This hidden folder is not part of the kernel source, but rather has to do with `git`.

`Documentation` – Pretty much self-explanatory. This is a collection of text files describing various aspects of the kernel, problems, "gotchas," and so on. There are many subdirectories under `Documentation/` for topics that require more extensive explanations. While the information here is generally useful, it also tends to be dated. That is, the documentation was initially written when a particular feature was added to the kernel, but it has not been kept up as the feature has evolved.

`arch` – All architecture-dependent code is contained in subdirectories of `arch`. Each architecture supported by the kernel has a directory under `arch` with its own subdirectory structure. The executable kernel image will end up in `arch/<architecture>/boot`. An environment variable in the makefile, `ARCH`, points to the appropriate target architecture directory. `arch/` is one of the two largest subtrees in the kernel, comprising just over 15,000 files.

`block` – The block layer. This is the code that optimizes transfers with block storage devices.

`crypto` – Code dealing with the cryptographic aspects of system security.

`drivers` – Device driver code. Under `drivers` is a large number of subdirectories for various devices and classes of device. This is the other major kernel subtree, with over 14,000 files.

`firmware` – Binary firmware files for devices that load firmware at boot time. Ultimately these files are to be moved to User Space, because they are, for the most part, proprietary and mixing them with the GPL'ed kernel code is "problematic."

`fs` – File systems. Under `fs` is a set of directories for each type of file system that Linux supports.

`include` – Header files. The most important subdirectory of `include/` is `linux/` that contains headers for the kernel APIs.

`init` – The basic initialization code.

`ipc` – Code to support Unix System 5 Inter-Process Communication mechanisms such as semaphores, message passing, and shared memory.

`kernel` – This is the heart of the matter. Most of the basic architecture-independent kernel code that does not fit in any other category is here. This includes things like the scheduler and basic interrupt handling.

`lib` – Several utility functions that are collected into a library.

`mm` – Memory management functions.

`net` – Network support. Subdirectories under `net` contain code supporting various networking protocols.

`samples` – Another relatively new addition, this has sample code showing how to manage some of the kernel's internal data structures.

`scripts` – Text files and shell scripts that support the configuration and build process.

`security` – Offers alternative security models for the kernel

`sound` – Support for sound cards and the Advanced Linux Sound Architecture.

tools – A performance monitor for the kernel.

`usr` – Mostly assembly code that sets up linker segments for an initial RAM disk.

`virt` – A virtual environment for the kernel.

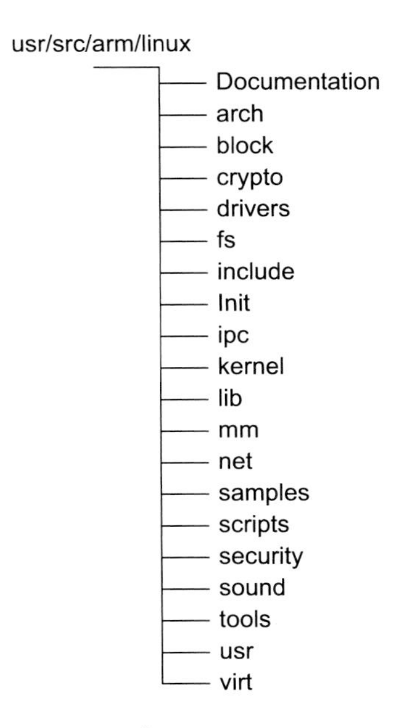


**Figure 12.3**
The kernel source tree.

## Kernel Makefile

`kernel/kernel` contains a standard makefile, `Makefile`, with a very large number of make targets. By default, the kernel is built for the architecture on which the makefile is running, which in the vast majority of cases is some variant of the x86. In our case, we want to cross-compile the kernel for our ARM target board.

As noted above, there is an environment variable, `ARCH`, that can be set to select the architecture to build for. This can be set on the command line that invokes `make` as in:

```
make ARCH=arm
```

Or, what I prefer to do, is edit the makefile to permanently set the value of `ARCH`. Open the makefile in `kernel/kernel` with your favorite editor,[1] and go to line 195, which initially reads:

```
ARCH        ?= $(SUBARCH)
```

Change it to read:

```
ARCH        ?= arm
```

The next line defines the variable `CROSS_COMPILE`. This identifies a cross-tool chain by specifying a prefix to be added to the tool name. Our ARM cross-tool chain is identified by the prefix `arm-linux-`. You can enter that here or, optionally, specify it as part of the configuration process. Save the file and exit from the editor.

Now execute `make help`. You will see a very long list of make targets in several categories. Our specific interest at the moment is the category Architecture specific targets (arm):. Under that is a large number of target names that end with "_defconfig." These make default configuration files for various ARM-based boards. When building a new kernel, it is always good practice to start with a known good configuration. Interestingly, the BeagleBone is not among this list of default configurations. Execute this command:

```
cp ../configs/beaglebone arch/arm/configs/beaglebone_defconfig
```

Now execute `make beaglebone_defconfig`. You will likely see a number of "errors" about "recursive dependency detected!" According to the forum at `beagleboard.org`, these are innocuous. The default configuration file `.config` is created.

## Configuring the Kernel: make config, menuconfig, xconfig

The process of building a kernel begins by invoking one of the make targets that carry out the configuration process. `make config` starts a text-based script that sequentially steps you

[1] There is not much point in using Eclipse for configuring and building the kernel. We will not be editing any source files, and the kernel has its own graphical configuration tools.

through each configuration option. For each option you have either three or four choices. The three choices are: "y" (yes), "n" (no), and "?" (ask for help). The default choice is shown in upper case.

Some options have a fourth choice, "m," which means build this feature as a loadable kernel module rather than build it into the kernel image. Kernel modules are a way of dynamically extending the kernel, and are particularly useful for things like device drivers.

Fig. 12.4 shows an excerpt from the `make config` dialog.

```
*
* Linux/arm 3.8.13 Kernel Configuration
*
*
* General setup
*
Prompt for development and/or incomplete code/drivers (EXPERIMENTAL) [Y/n/?]
Cross-compiler tool prefix (CROSS_COMPILE) []
Local version - append to kernel release (LOCALVERSION) []
Automatically append version information to the version string (LOCALVERSION_AUTO)
 [Y/n/?]
Kernel compression mode
  1. Gzip (KERNEL_GZIP)
  2. LZMA (KERNEL_LZMA)
  3. XZ (KERNEL_XZ)
> 4. LZO (KERNEL_LZO)
choice[1-4?]:
Default hostname (DEFAULT HOSTNAME) [(none)] █
```

**Figure 12.4**
Make config dialog.

Most options include help text that is genuinely "helpful" (see Fig. 12.5).

```
Default hostname (DEFAULT_HOSTNAME) [(none)] ?

CONFIG_DEFAULT_HOSTNAME:

This option determines the default system hostname before userspace
calls sethostname(2). The kernel traditionally uses "(none)" here,
but you may wish to use a different default here to make a minimal
system more usable with less configuration.

Symbol: DEFAULT_HOSTNAME [=(none)]
Type  : string
Prompt: Default hostname
  Defined at init/Kconfig:208
  Location:
    -> General setup


Default hostname (DEFAULT_HOSTNAME) [(none)] █
```

**Figure 12.5**
Kernel configuration help text.

The problem with `make config` is that it is just downright tedious. Typically, you will only be changing a very few options and leaving the rest in their default state. But `make config` forces you to step through each and every one, and contemporary kernels have well over 2000 options. Personally, I've never used `make config` and I honestly wonder why it is still in the kernel package.

`make menuconfig`, based on the `ncurses` library, brings up the pseudo-graphical screen shown in Fig. 12.6. Here, the configuration options are grouped into categories and you only need to visit the categories of options you need to change. The interface is well explained and reasonably intuitive. But since it is not a true graphical program, the mouse does not work. The same help text is available as with `make config`. When you exit the main menu, you are given the option of saving the new configuration.

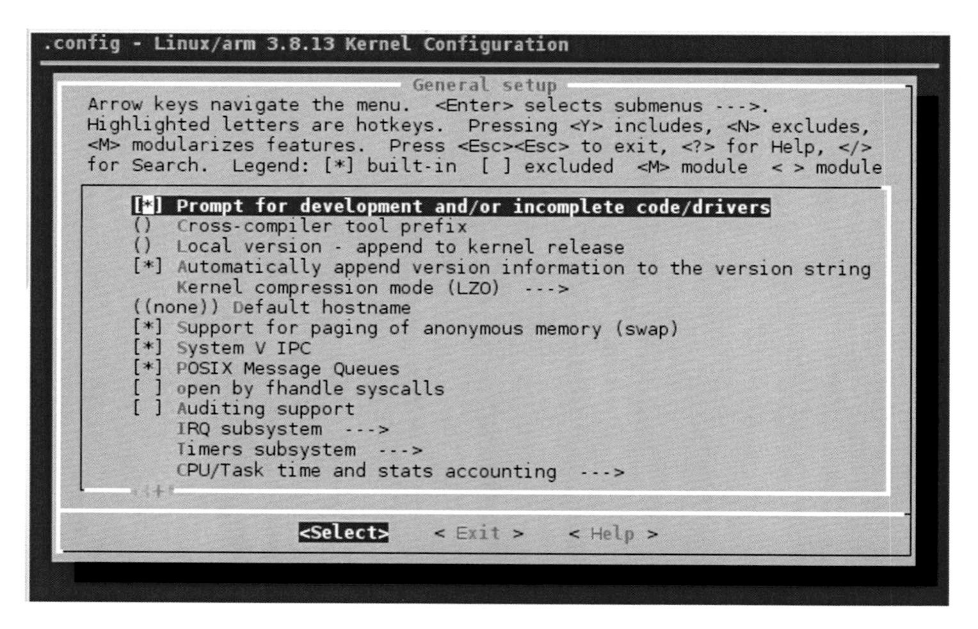


**Figure 12.6**
Make menuconfig.

While most true Linux hackers seem to prefer `menuconfig`, my choice for overall ease of use is `make xconfig`. This brings up an X Windows-based menu, as shown in Fig. 12.7. Now you can see all the option categories at once, and navigate with the mouse. Of course you must be running X Windows to use this option, and you must have the g++, Qt graphics library, and Qt development tools packages installed.

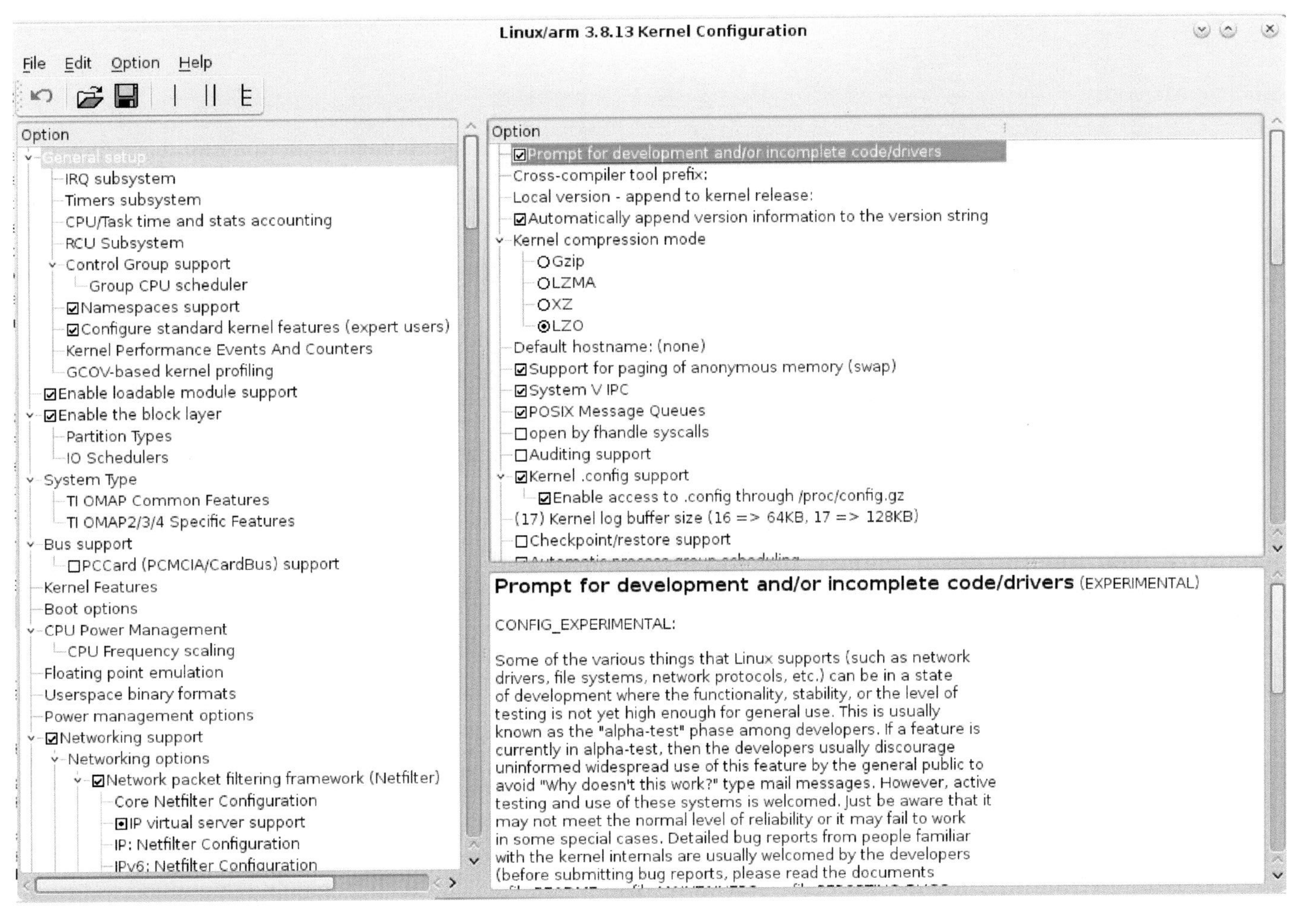


**Figure 12.7**
Make xconfig.

In its default display mode, `xconfig` displays a tree of configuration categories on the left. Selecting one of these brings up the set of options for that category in the upper right window. Selecting an option then displays help for that option in the lower right window. Fig. 12.7 starts with the general setup menu. Most of the options are presented as check boxes. Clicking on the box alternates between checked (selected) and unchecked (not selected). These are termed *binary* options. Options that may be built as kernel modules have a third value, a dot to indicate that the feature will be built as a module. These are *tri-state* options. An example of that is `Bus support`. Click the box next to `PCCard (PCMCIA/CardBus) support`, and it changes to a dot. Click it again and it changes to a check mark. Click it again to leave the box unchecked.

Some options have a set of allowable values other than yes or no, and are represented by *radio buttons*. Select `Kernel features` to see a couple of examples of this: `Memory split` and `Preemption Model`. Some options take numeric values. An example of a numeric option is back up under `General setup, kernel log` buffer `size`. It is not immediately obvious how you change a numeric value. Double-clicking the option opens a dialog box between the two right-hand windows.

Options that are not available, because some other option was not selected, simply do not show up. Again, the `PC Card` option serves as an example. With the box unchecked, the upper right-hand window is empty. Now click the check box, and the appropriate options show up in the upper right.

**Try it out**

The best way to get familiar with kernel configuration options is to fire up xconfig and see what is there. So...

```
cd kernel/kernel
make xconfig
```

After a bit of program and file building, the menu shown in Fig. 13.7 will appear. Just browse through the submenus and their various subsubmenus to get a feel for the flexibility and richness of features in the Linux kernel. Read the help descriptions.

Most of the critical configuration options should be set correctly in the default .config file, but there is one option that is not yet set. Under General setup is an option called Cross-compiler tool prefix. This is a string option that must be set to "arm-linux- ." Be sure that NFS client support and Root file system on NFS are selected under Network File Systems.

The default configuration enables any number of options that you do not really need, at least not initially. The BeagleBone Black has no built-in wireless support, so Bluetooth and anything called "Wireless" can be safely removed.

For the time being, leave the configuration menu up while you read the next section.

`make xconfig` gives you additional flexibility in saving configurations. Instead of always saving it to the standard `.config` file, you can save it to a named file of your choice using the `File->Save As` option, and later load that file using `File->Load` for further modification or to make it the new default configuration.

### *Xconfig Options*

The `xconfig` menu has several useful options in the **Option** menu. **Show Name** adds a column to the upper right panel that displays the name of each configuration variable. **Show Range** displays the range and current value of binary and tri-state options. In my estimation, this is only marginally useful. Another marginally useful option is **Show Data**. For binary and tri-state options it pretty much duplicates **Show Range**. It can be helpful for numeric and text options.

**Show All Options** is largely self-explanatory. It displays in gray all of the options that cannot be selected because some condition is not satisfied. Suppose you are pretty sure some option exists, but you cannot find it. Turn on **Show All Options**, then use **Edit -> Find** to search for the name.

There are some options that you are simply not allowed to change. These are preselected based on the architecture you are building for. These options lack a "prompt" that lets you edit them. With **Show All Options** turned on, you will find most of these options at the top of the menu. If you select **Show Prompt Options**, you will only see those options that you can change, provided the dependencies are met.

When you are finished, close the xconfig menu. You will be asked if you want to save or discard the changes.

### *.config File*

It is probably worth mentioning that true kernel hackers tend to prefer `make menuconfig` over `make xconfig`. The end result of the configuration process, whichever one you choose to use, is a file called `.config` containing all of the configuration variables. Listing 12.1 is an excerpt. The first thing to notice is the comment at the top:

```
# Automatically generated file; DO NOT EDIT
```

It is considered "bad form" to manually edit files that have been created by a tool. There is a high probability that you will mess something up.[2]

[2] OK, I've manually edited configuration files. Everybody does it. Just understand that in doing so, you are taking a risk.

The options that are not selected remain in the file but are "commented out." The selected options get the appropriate value, "y" for features to be built into the kernel image, "m" for modules, a number or an element from a list. `.config` is included in the Makefile, where it controls what gets built and how. Early in the build process `.config` is converted into a C header file, `include/linux/autoconf.h`, so that the configuration variables can be used to control conditional compilation.

```
#
# Automatically generated file; DO NOT EDIT.
# Linux/arm 3.8.13 Kernel Configuration
#
CONFIG_ARM=y
CONFIG_SYS_SUPPORTS_APM_EMULATION=y
CONFIG_GENERIC_GPIO=y
CONFIG_HAVE_PROC_CPU=y
CONFIG_STACKTRACE_SUPPORT=y
CONFIG_LOCKDEP_SUPPORT=y
CONFIG_TRACE_IRQFLAGS_SUPPORT=y
CONFIG_RWSEM_GENERIC_SPINLOCK=y
CONFIG_ARCH_HAS_CPUFREQ=y
CONFIG_GENERIC_HWEIGHT=y
CONFIG_GENERIC_CALIBRATE_DELAY=y
CONFIG_NEED_DMA_MAP_STATE=y
CONFIG_VECTORS_BASE=0xffff0000
CONFIG_ARM_PATCH_PHYS_VIRT=y
CONFIG_GENERIC_BUG=y
CONFIG_DEFCONFIG_LIST="/lib/modules/$UNAME_RELEASE/.config"
CONFIG_HAVE_IRQ_WORK=y
CONFIG_IRQ_WORK=y
CONFIG_BUILDTIME_EXTABLE_SORT=y
```

**Listing 12.1**
Excerpt of `.config`.

## A Problem

It turns out that the kernel we just patched and configured will not build. The file `arch/arm/kernel/return_address.c` does not compile, because a function it contains is also defined as an inline in `arch/arm/include/asm/ftrace.h`. Internet research revealed that the problem was that the kernel would not build under gcc version 6, and the solution was to drop back to gcc version 4.9.

So I built a gcc 4.9 tool chain, and had exactly the same problem. If you look at `ftrace.h` you can see the problem. There is a conditional based on a couple of configuration variables that ends up declaring the function `return_address()` as inline rather than a simple prototype. The real problem is that these configuration variables are controlled by `depends` and `selects`, and that makes it very difficult to change them using `make xconfig`.

My solution then was to edit `ftrace.h` to eliminate the conditional entirely by commenting it out, and just leave the function prototype. Listing 12.2 highlights my edits. The lines that are changed are emboldened.

```
// #if defined(CONFIG_FRAME_POINTER) && !defined(CONFIG_ARM_UNWIND)
/*
 * return_address uses walk_stackframe to do it's work.  If both
 * CONFIG_FRAME_POINTER=y and CONFIG_ARM_UNWIND=y walk_stackframe uses un wind
 * information.  For this to work in the function tracer many functions would
 * have to be marked with __notrace.  So for now just depend on
 * !CONFIG_ARM_UNWIND.
 */

void *return_address(unsigned int);

//#else

//static inline void *return_address(unsigned int level)
//{
//        return NULL;
//}

//#endif
```

**Listing 12.2**
ftrace.h.

I believe this is the first time I've had to edit a kernel source file in order to get the kernel to build. I do not take this lightly, and neither should you. Changing kernel code is not something you do on a whim, because there is a very real possibility your changes will create some other problem down the road. In this case it seems to work.

## *Building the Kernel*

The actual process of building the kernel varies a little depending on what you are building it for. Our primary objective is to build a kernel for the target board, so I'll describe that first. Then I'll describe the alternate process of building a new kernel for your workstation if you are so inclined.

The first two steps below can be executed as a normal user.

`make clean`. Deletes all intermediate files created by a previous build. This ensures that *everything* gets built with the current configuration options. You will find that virtually all Linux makefiles have a `clean` target. Strictly speaking, if you are building for the first time, there are no intermediate files to clean up, so you do not have to run `make clean`.

`make` This is the heart of the matter. This builds the executable kernel image and all kernel modules. Not surprisingly, this takes a while. The resulting compressed kernel image is `arch/$(ARCH)/boot/zImage`.

The following step requires root user privileges.

`make modules_install INSTALL_MOD_PATH=/export/rootfs`. Copies the modules to `$(INSTALL_MOD_PATH)/lib/modules/<kernel_version>` where `<kernel_version>` is the string identifying the specific kernel you are building. Note that when building for the target board, you must explicitly specify the root file system. Otherwise, the modules will go into `/lib/modules` of your workstation's file system. Strictly speaking, this step is not necessary in this case, because none of the modules are required to boot the kernel.

That is it. Note, incidentally, that the build process is recursive. Every subdirectory in the kernel source tree has its own Makefile dealing with the source files in that directory. The top level Makefile recursively invokes all of the sub Makefiles.

The process for building a kernel for your workstation differs in a couple of details. Do `make modules_install` without the `INSTALL_MOD_PATH` argument, and then execute this step, also as root:

`make install`. This does several things. It copies the kernel executable, called `vmlinuz` for x86 builds, to `/boot` along with `System.map`, the linker map file. It adds a new entry to `/boot/grub/grub.conf`, so the GRand Unified Bootloader can offer the new kernel as a boot option. Finally, it creates an *initial ramdisk*, initrd, in `/boot`.

### Workstation Digression

Before we move on to loading and testing our new kernel, the last two items mentioned above deserve a passing explanation. Most workstation installations these days incorporate a boot loader called GRUB (GRand Unified Bootloader) to select the specific kernel or alternate operating system to boot. There is a very good reason for having the ability to boot multiple kernel images. Suppose you build a new kernel and it fails to boot properly. You can always go back and boot a known working image, and then try to figure out what went wrong in your new one. If you are curious, have a look at the file `/boot/grub2/grub.cfg` on your workstation. You will need to be root user to get into the grub2 directory.

Most Linux kernels are set up to use `initrd`, which is short for initial ramdisk. An initial ramdisk is a very small Linux file system loaded into RAM by the boot loader and mounted as the kernel boots, before the main root file system is mounted. The usual reason for using initrd is that some kernel modules need to be loaded before mounting the root partition. Usually, these modules are required to support the file system used by the root partition, ext3 for example, or perhaps the controller that the hard drive is attached to, such as SCSI or RAID.

## Booting the New Kernel

Now that we have a new kernel, how do we test it? We could, of course, load the new image into flash, either in addition to, or in place of, the kernel image presently there.

The other alternative is to boot the new image over the network using TFTP. This is particularly advantageous in a development environment, because we can quickly test a new kernel image without the time-consuming process of burning flash.

Move `arch/arm/boot/zImage` to `/var/lib/tftpboot`, or wherever else you chose as the TFTP directory. Probably the simplest way to accomplish this is with a minor change to `uEnv.txt`. Copy `uEnv.txt` to `uEnvnet.txt`, and open with an editor. Note the line near the bottom that starts out `netboot=`. This command loads the kernel image into memory over TFTP. This is the equivalent of `loadkernel` that is presently used to load the kernel image from eMMC flash. So, in the `uenvcmd`, just replace `loadkernel` with `netboot`.

As we did back in Chapter 6, The hardware when bringing up the BBB, save `uEnvnet.txt` to the boot partition of the eMMC. Boot the board into the u-boot prompt, change `bootenv` to `uEnvnet.txt`, and execute `boot`. After the kernel boots, execute the command `uname -a` and note that the kernel image has a timestamp corresponding to when you built it.

Congratulations! You can pat yourself on the back. You are now a Linux hacker.

## Behind the Scenes: What's Really Happening

The information in this section is not essential to the process of building a kernel, and you are free to ignore it for now. But when you reach the point of developing device drivers or other kernel enhancements (or perhaps hacking the kernel itself), you will need to modify the files that control the configuration process.

All the information in the configuration menus is provided by a set of text files named `Kconfig` that are scattered throughout the source tree. These are script files written in *Config Language*, which looks suspiciously like a shell scripting language but isn't exactly. The main `Kconfig` file is located in `linux/arch/$(ARCH)`, where `ARCH` is the variable in `Makefile` identifying the base architecture.

Go to `kernel/arch/arm` and open `Kconfig`. Find the line that reads `menu "System Type"` down around line 249. Compare the structure of the file with the configuration menu starting at System Type, and the pattern should become fairly clear. Each configuration option is identified by the keyword `config` followed by the option's symbolic name, for example `config MMU`. Each of these gets turned into a Makefile variable, or macro, such as `CONFIG_MMU`. The Makefiles then use these variables to determine which components to include in the kernel and to pass `#define` symbols to the source code.

The option type is indicated by one of the following keywords:

- `bool` – The option has two values, "y" or "n."
- `tristate` – The option has three values, "y," "n," and "m."
- `choice` – The option has one of the listed values.
- `int` – The option takes an integer value.

The type keyword usually includes a prompt, which is the text displayed in the configuration menu. Alternatively, the `prompt` keyword specifies the prompt. There are also "hidden" options that have no displayable prompt. At the top of `Kconfig` is a set of options that do not show up in the menu, because no prompt is specified. These are the options you see when you select **Show All Options** in xconfig.

Other keywords within a configuration entry include `help` to specify the help text, and `depends`, which means that this option is only valid, or visible, if some other option that it depends on is selected. There is also `select`, which says that if this option is selected, then of necessity some other option must also be turned on.

Just above the `menu "System Type"` line is the line:

```
source "kernel/Kconfig.freezer"
```

The `source` keyword is how the configuration menu is extended to incorporate additional, modular features. It is effectively the same thing as `#include` in C. You will find `source` lines scattered throughout the main `Kconfig` file.

Config Language is actually much more extensive than this simple example would suggest. For more detail, look at `linux/Documentation/kbuild/kconfig-language.txt`.

In the next chapter, we will look at *integrated build environments*, open source projects that attempt to do everything for you to build an embedded Linux system.

## Resources

For more details on the process of configuring and building a kernel, look at the files in `/usr/src/arm/linux/Documentation/kbuild`. Additionally, the following HOW-TOs at www.tldp.org may be of interest:

`Config-HOWTO` This HOWTO is primarily concerned with how you configure your system once it is built.

`Kernel-HOWTO` Provides additional information on the topics covered in this chapter.

www.linuxfromscratch.org/ – This is a project that provides you with step-by-step instructions for building your own custom Linux system, entirely from source code. There is a subproject that deals with cross-building Linux for embedded environments.

lxr.linux.no – This is probably my favorite web resource for the Linux kernel. It is the kernel source tree hyperlinked and cross-referenced. Virtually all versions of the kernel are available. You can look up any symbol in the kernel, see where it is declared, and where it is used. You can step through a hierarchy of function calls to see what is really going on. It is just great.

CHAPTER 13

# Integrated build environments

## Chapter Outline



*It's not an embedded Linux distribution—it creates a custom one for you*
***From the Yocto Project website***

In this chapter, we will explore a growing trend in embedded Linux development—the use of integrated build environments to build all or part of an embedded Linux distribution.

## The Problem

There are a number of elements that go into an embedded development environment. These include among others:

- Cross-tool chain
  - Editor
  - Compiler
  - Linker
  - Debugger
  - Libraries
- Bootloader
- Linux kernel

Linux for Embedded and Real-time Applications.
DOI: http://dx.doi.org/10.1016/B978-0-12-811277-9.00013-4

- Root file system
- Application program(s)

Getting all of these pieces to "play together" nicely is a non-trivial exercise. Not to mention licensing issues. As we saw in Chapter 1, The embedded and real-time space, open source software is released under a variety of licenses with different provisions. It is important to be sure the various licenses also play well with one another.

Where do all these pieces come from? Ultimately they come from various places around the Internet. That is all well and good, but how do you find them?

The promise of integrated build environments is that they bring everything together in one place. Someone else has done the work of finding all these pieces on the net, and verifying that they all work together. One feature all of these projects have in common, however they are implemented, is that once you finish the configuration process, they all go out on the Internet and download the required sources and build the components you have specified. One consequence of this is that the initial build typically takes a *long* time. Subsequent modifications typically do not take very long, because only a small amount of code is being downloaded and rebuilt. Other features they all share: they use the same configuration tools as the kernel, and they have plug-ins for Eclipse.

We will take a tour of three different projects. The websites for these projects are listed in the "Resources" section. Feel free to follow along, or not, on your own workstation as you choose.

## Buildroot

Buildroot is probably the simplest of the integrated build environments. It is a set of Makefiles, scripts, and patches that can build all the elements of an embedded Linux system as listed above, or any combination of them. Buildroot can automatically build the required cross-compilation tool chain, create a root file system, compile a Linux kernel image, and build a bootloader for the targeted embedded system, or it can perform any independent combination of these steps.

Unlike the kernel, buildroot's `make help` does not list the default configurations it can build. Instead there's a `make list-defconfigs` target and indeed there is a default configuration for the BeagleBone. So execute `make beaglebone_defconfig`. Next, execute make xconfig. The menu is shown in Fig. 13.1.

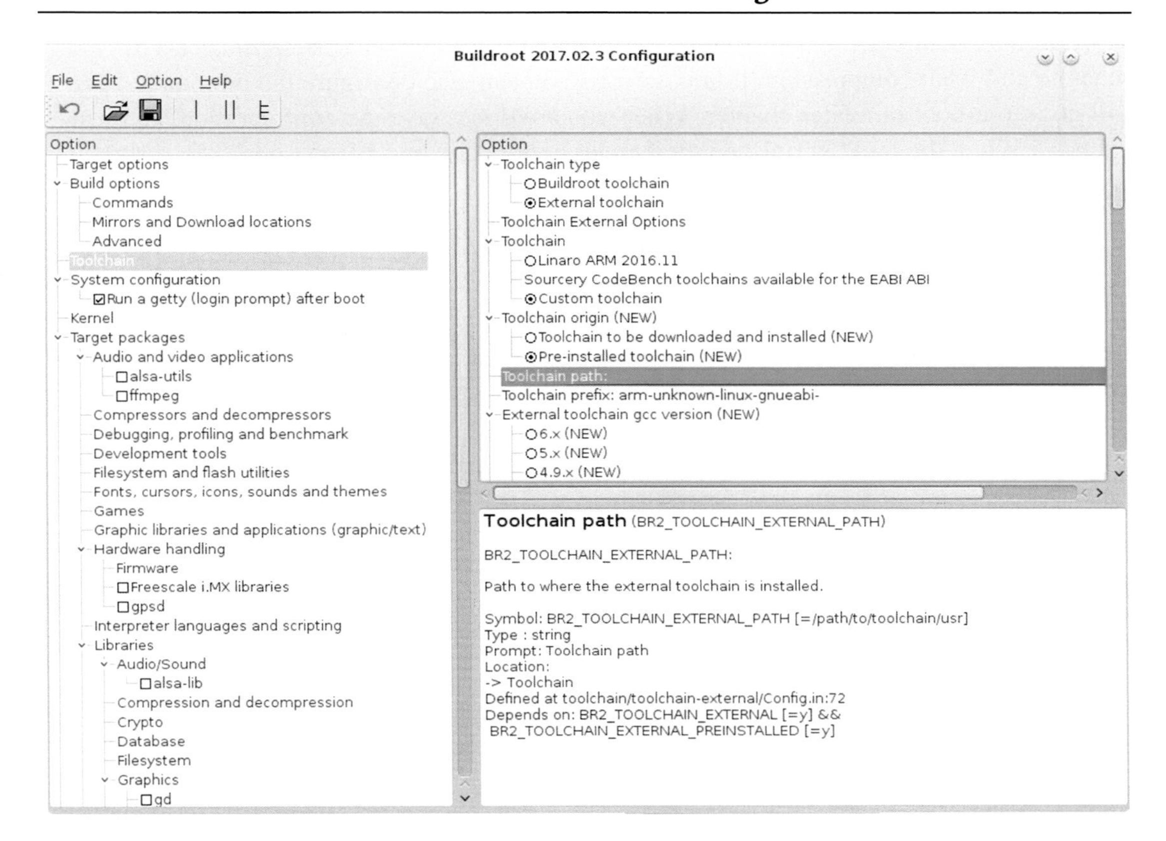


**Figure 13.1**
Buildroot xconfig menu.

Buildroot offers the option of using an existing tool chain, rather than building a new one from source. Since we have a tool chain, that would seem like a sensible choice. In the `Toolchain` category, select `Custom toolchain` and `Pre-installed toolchain`. You can set the `Toolchain path` to, since the tool chain is in your `PATH`. Set the `Toolchain prefix` to `arm-unknown-linux-gnueabi`.

The `External toolchain gcc version` is most likely `4.9.x`. In the previous chapter we built kernel version 3.8.13, so you would expect to select `3.8.x` for `External toolchain kernel headers series`. It turns out though, that pretty much everything earlier than 4.1.x is no longer supported and if you select it, the build fails. So, even though we already have a Linux kernel source tree, if you want to use Buildroot, you will have to select one of their kernels.

Much of the rest of the Buildroot configuration has to do with selecting packages to install on the target. You will not need to select much here, because most utilities will be provided by Busybox, which we will go into in the next chapter.

Down at the bottom of the menu, you can select what kind of file system image(s) to generate and what compression, if any, to use. You can also configure the bootloader. We will go into u-boot in a later chapter. When you are done, save the configuration, and just run `make`. When the build is finished the kernel, bootloader, and file system images show up in `output/images/`.

## Open Embedded

While Open Embedded, simply called OE for short, was probably the first attempt at a true integrated build environment, its original implementation, now called OpenEmbedded-Classic, has been substantially overhauled. The project began by merging the efforts of several projects aimed at supporting personal digital assistants running Linux including OpenZaurus, Familiar Linux, and OpenSIMpad.

The current incarnation of OE, called OpenEmbedded-Core, or OE-Core for short, resulted from a merger with the Yocto Project that we will see a little later. Even though OE-Core remains an active project, much of the ongoing development is being done under the auspices of the Yocto Project. In particular, that is where the most current documentation resides.

> WARNING: OE-Core takes a *lot* of disk space. The Yocto Project suggests that you need 50 GB of free space, which of course must all be in one partition. So, if you intend to actually build something with OE-Core or Yocto, be sure you have enough free space.

OE-Core is based on a build tool called BitBake written in Python. BitBake's operation can be represented graphically, as shown in Fig. 13.2. It interprets, or parses if you prefer, *recipes* and *configuration files* to determine what needs to be built and how. Then it *fetches* the necessary source code over the network. The output is a set of *packages* and file system *images*.

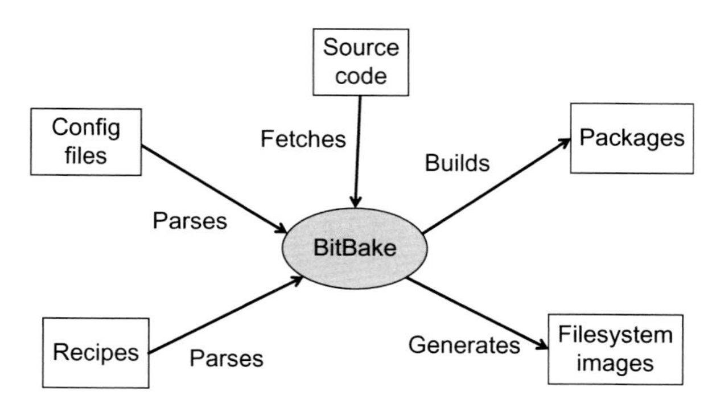


**Figure 13.2**
BitBake operation.

BitBake is now a sub-project of the Yocto Project.

OE-Core itself then is a base layer of recipes, classes, and associated files that is meant to be common among many different OpenEmbedded-derived systems.

There is a *Getting Started* guide in the documentation section of the OE web site. What follows is derived from that. Consult the guide for more details.

The first step is to make sure you have all of the necessary packages installed. Here is a complete list of the packages required when installing under CentOS:

- `bzip2`
- `chrpath`
- `cpp`
- `diffstat`
- `diffutils`
- `gawk`
- `gcc`
- `gcc-c++`
- `git`
- `glibc-devel`
- `gzip`
- `make`
- `patch`
- `perl`
- `python`
- `SDL-devel`
- `tar`
- `texinfo`
- `unzip`
- `wget`
- `xterm`

The list for other distributions is similar. In reality, most of these were installed during the original CentOS installation. Here is the command for the packages I had to install (run as root of course):

```
yum install texinfo chrpath SDL-devel
```

It turns out that the version of python in the CentOS repo is too old for OE-Core. The upshot is you have to download and install python3. Go to the python web site, download the latest, and consult the README file for instructions on installation.

To install OE-Core, execute these commands from your home directory:

```
git clone git://git.openembedded.org/openembedded-core oe-core
cd oe-core
git clone git://git.openembedded.org/bitbake bitbake
git checkout pyro
cd bitbake
git checkout 1.34
cd..
```

The first thing to notice is that OE-core is substantially larger than buildroot: about 123 MB vs 17 MB. The next step is to initialize the build environment by executing the `oe-init-build-env` script like this:

```
source oe-init-build-env [build_dir]
```

Normally, the shell spawns a new process every time it executes a command or script. The `source` command causes the script to be executed in the current shell environment. The optional argument is the subdirectory of oe-core/ that holds all the files created during the build. The default is `build/`. This script creates a couple of configuration files with default values, creates and changes to the `build/` directory, and adds some entries to your `PATH` environment variable. Execute `echo $PATH`, and note three new entries at the beginning. If you intend to use OE_Core for real, you might want to add these to the `PATH` definition in your `.bash_profile` file. Otherwise, you will have to run the `oe-init-build-env` script every time you want to do something with OE-Core.

One of the files just created is `build/conf/local.conf`. This provides general configuration for the images you build. Let us have a look. The file is well-commented, describing each variable. The first is the `MACHINE` for which to build images. All of the possibilities are listed as comments. There are several QEMU targets, as well as some specific hardware targets. The default is `qemux86`, the 32-bit Intel architecture running under QEMU, an open source multi-platform emulator.

Next, some directories are specified. Default values are subdirectories of `TOPDIR`, the `build/` directory:

- `DL_DIR`—where downloaded source packages are stored
- `SSTATE_DIR`—where *shared state* files are stored. This speeds up subsequent builds
- `TMPDIR`—where build output is stored

We can specify which package management format to use with the `PACKAGE_CLASSES` variable. Unless you are building for a "debian flavor" distro, you should probably select rpm.

Save `local.conf`, and we are ready to build a cross-tool chain and a set of images. OE-Core provides a set of predefined targets that we can use to build our own images. A couple of these were listed by `oe-init-build-env`. To see the complete set of image targets execute:

```
ls ../meta*/recipes*/images/*.bb
```

Now to build the `core-image-minimal` image, we simply execute:

```
bitbake core-image-minimal
```

### *OE Metadata*

While the image is "baking," it is an opportunity to explore OE metadata. The *metadata* that drives BitBake consists of recipes and configuration files. Recipes get the file extension `*.bb`, and configuration files get the extension `.conf`. All of this metadata resides in the subdirectory `meta/`. Although it is not obvious from the layout of `meta/`, the recipes are divided into a hierarchy of four categories:

- Image
- Task
- Package
- Class

The motivation for the hierarchical organization is that recipes in lower levels can be easily reused.

Image is the top level of the hierarchy. An image recipe defines what goes into a specific root file system image. It lists the tasks that make up the image. Have a look at `meta/recipes-core/images/core-image-minimal.bb`. It installs one task, `packagegroup-core-boot`, and possibly others defined in `CORE_IMAGE_EXTRA_INSTALL`. It also inherits a class called `core-image`. Class files get the extension `.bbclass`, and reside in `meta/classes/`.

Task recipes typically represent aggregations of packages. In fact, many of them are named `packagegroup-*`. For example take a look at `packagegroup-core-boot.bb`. It claims to be "The minimal set of packages required to boot the system." Among other things, it inherits a class called `packagegroup`.

A package recipe manages the specific needs of an individual package. Package recipes typically have a version number appended to the name. `meta/recipes-core/sysvinit/sysvinit_2.88dsf.bb` is a typical example. It identifies where the package is found on the net, the license terms, the required files, and so on.

Class recipes have a peer-to-peer relationship with the other recipes, rather than a hierarchical one. They are used to factor out common recipe elements that can then be reused, through the inherit command. The most frequent use of classes is to inherit functions or portions of recipes for commonly used build tools like the GNU autotools. Have a look at `meta/classes/core-image.bbclass`. It in turn inherits `image.bbclass`.

### Testing the New Image: Quick EMUlator

How do we test our newly created image? OE-Core includes a tool called QEMU, which stands for Quick EMUlator. QEMU is an open source project that emulates several different machine architectures. The image we just built is for the x86 running under QEMU, and is named qemux86. So to test it, all we have to say is:

```
runqemu qemux86
```

### Personal Observations

Personally, I do not care much for Open Embedded, or the Yocto Project that we will look at next. They are simply too big, too complex, too obscure. The learning curve is steeper than for Linux itself. To quote the introductory paper listed in Resources:

> [I]t has a considerable learning curve and the source of build errors may not be obvious. Development teams that are unfamiliar with OE should allow at least several weeks to attain moderate proficiency.

It is not the least bit obvious how you would modify any of the over 4000 files in the `meta/` directory to effect any meaningful change in your root file system image. As far as I can tell, there is no graphical configuration tool.

## Yocto Project

The Yocto Project is a collaborative effort of the Linux Foundation. The goal is to produce tools and processes that will facilitate creating Linux distributions for embedded software that are independent of the underlying architecture of the embedded software itself. The project was announced by the Linux Foundation in 2010, and in March 2011, the project aligned itself with Open Embedded to create The OpenEmbedded-Core Project. Yocto's reference implementation of OE-Core is called Poky (pronounced PAH-kee, not POH-kee).

Yocto appears to be somewhat more comprehensive than OE-Core. Fig. 13.3 shows the Yocto workflow.

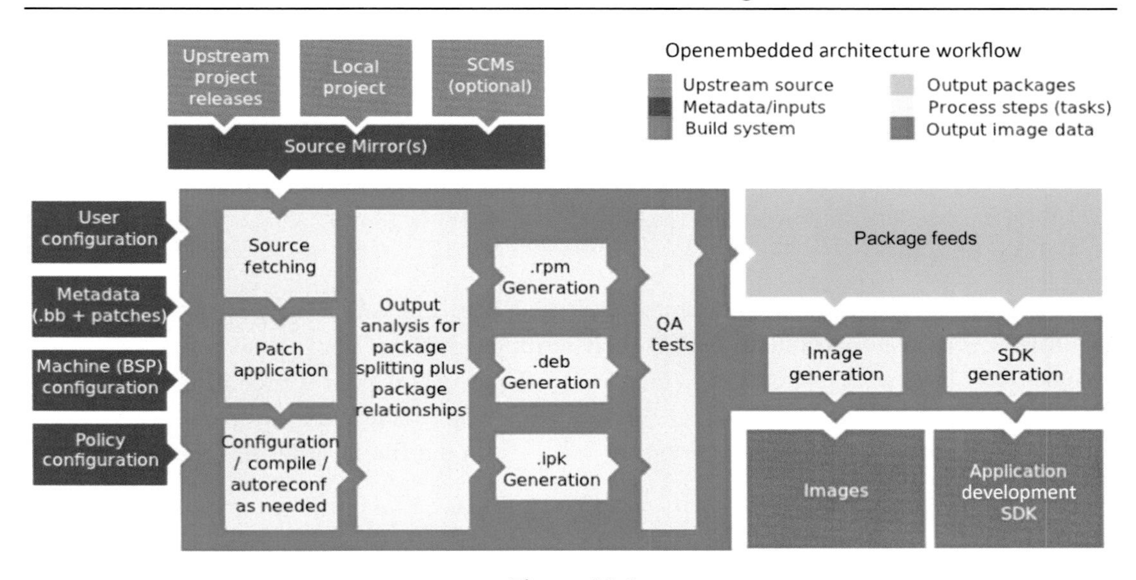


**Figure 13.3**
Yocto Project development environment.

The Poky build system is available in two forms: as a git repository or as a single Bzip file. `cd` to your home directory and do one of the following:

```
git clone -b pyro git://git.yoctoproject.org/poky.git
git checkout pyro
```

Download `poky-pyro-17.0.0.tar.bz2` from the Yocto Project downloads page and untar it. For simplicity, rename the directory `poky-pyro-17.0.0` as just `poky`.

This time we will build an image for the BBB. That requires editing `poky/build/local.conf` to change the target machine. Find the line `#MACHINE ?= "beaglebone"` and remove the leading hash mark. Save the file.

The process for building an image is very much like OE-Core. We first have to initialize the build environment:

```
source oe-init-build-env
```

Let us build something just a little more ambitious this time:

```
bitbake core-image-full-cmdline
```

### *Application Development with Yocto: The Software Development Kit and Eclipse*

So far we have talked about building bootloaders, kernels, root file systems, and tool chains, but we have not said anything about application development, which is probably your primary objective, right? The Yocto Project includes a Software Development Kit (SDK) intended to ease application development.

The major features of the SDK are:

- Cross-tool chain with compiler, debugger, QEMU, and other miscellaneous tools
- Libraries, headers, and symbols
- Eclipse plug-in

There is a separate SDK for each architecture that Yocto supports. These are distributed in the form of very large (around 200 MB) shell scripts. Go to the Yocto downloads page and navigate to `releases/yocto/yocto-2.3//toolchain`. Then `i686` if your workstation is 32 bits or `x86_64` if it is 64 bits. Download `poky-glibc-x86_64-core-image-minimal-cortexa8hf-neon-toolchain-ext-2.3.sh`. You will need to run `chmod +x` on the file in order to execute it.

The default install location is `/opt/poky/2.3`. You can pass a `-d` option to the script to specify a different location. I chose to install it under `~/poky/2.3`. Like Poky itself, there is a configuration script that must be sourced to set up the environment. Run this command:

```
source environment-setup-cortexa8hf-neon-poky-linux-gnueabi
```

The Yocto Eclipse plug-in only works with the Neon version of Eclipse (notice the neon in the script file name). Installation is a bit different with Neon than with previous versions:

1. Go to the Eclipse web site and download the "installer" tarball
2. Expand that
3. `cd eclipse-installer`
4. Run `eclipse-inst`
5. Select `Eclipse for C/C++ Developers`
6. Specify the `Installation Folder`, and
7. Click `Install`.

Once you have Eclipse Neon running, there are several Eclipse plug-ins we need to install before installing the Yocto plug-in:

1. Select `Install New Software` from the `Help` menu
2. In the `Work with` dropdown, select `Neon—http://download.eclipse.org/releases/neon`
3. Expand `Linux tools`
4. Check the boxes for both entries: `C/C++ Remote (Over TCF/TE) Run/Debug Launcher` and `TM terminal`
5. Expand Mobile and Device Development and check the following:
    - `C/C++ Remote (Over TCF/TE) Run/Debug Launcher`
    - `Remote System Explorer User Actions`
    - `TCF Remote System Explorer add-in`

- `TCF Target Explorer`
- `TM Terminal`

6. Expand `Programming Languages` and select `C/C++ Development Tools SDK`
7. Click Next twice, accept the license terms and click Finish.

You are prompted to restart Eclipse for the changes to take effect. Not yet, because now we want to install the Yocto plug-in. Go back to `Install New Software`:

1. Click `Add...` in the `Work with` area
2. Enter `http://downloads.yoctoproject.org/releases/eclipse-plugin/2.3/neon` in the `Location` field, and provide some meaningful name in the `Name` field. Click `OK`
3. Check both of the entries that show up, click `Next` twice, accept the license, and click `Finish`. Interestingly enough, a `Security Warning` pops up that says the software is unsigned and its authenticity cannot be established. Go ahead and click `OK`.

Now you can restart Eclipse. You will find a new menu option, `Yocto Project Tools`. Also under `Window>Preferences` there is a dialog for `Yocto Project SDK`.

The *Yocto Project Software Development Kit (SDK) Developers' Guide 2.3*, available from the Documentation page of the Yocto Project website, has an excellent chapter on using the Eclipse plug-in. It in turn points to a wiki entry, with more details on setting up the plug-in.

### Personal Observations

Yocto is probably an improvement over OE-Core. At least it has some graphical tools. Nevertheless, like OE-Core, it is very big, very complex, and requires a steep learning curve. If you can invest the time in really learning it, then it is probably useful.

In the next chapter, we will tackle Busybox, the Swiss army knife of embedded Linux.

## Resources

www.buildroot.org—Website for the Buildroot project.
www.openembedded.org—Website for the Open Embedded project.
www.yoctoproject.org—Website for the Yocto Project.
http://downloads.yoctoproject.org—Downloads page for the Yocto Project.
www.python.org—Website for python.
www.embedded.com/design/prototyping-and-development/4218490/Open-Embedded--An-alternative-way-to-build-embedded-Linux-distributions—An introductory paper on Open Embedded Core. It is a little old, but still useful.
https://lwn.net/Articles/682540/—*Deciding Between Buildroot and Yocto* reports on a session at the 2016 Embedded Linux Conference that compared and contrasted Buildroot and the Yocto Project.

CHAPTER 14

# BusyBox and Linux initialization

## Chapter Outline



*Linux: the choice of a GNU generation*

***ksh@cis.ufl.edu put this on T-shirts in 1993***

Very often the biggest problem in an embedded environment is the lack of resources, specifically memory and storage space. As you have no doubt observed, either in the course of reading this book, or from other experience, Linux is *big*! The kernel itself is often in the range of two to four megabytes, and then there is the root file system, with its utility programs and configuration files. Granted, our BBB is running a full Debian distro, but still, we are limited to the 4 GB eMMC, or a microSD card.

In this chapter, we will look at a powerful tool for substantially reducing the overall "footprint" of Linux to make it fit in limited resource embedded devices. The other topic we will address in this chapter is User Space initialization, and specifically the question of how to get our thermostat application to start at boot up.

Linux for Embedded and Real-time Applications.
DOI: http://dx.doi.org/10.1016/B978-0-12-811277-9.00014-6

## Introducing BusyBox

Even if your embedded device is "headless," that is it has no screen and/or keyboard in the usual sense for user interaction, you still need a minimal set of command line utilities. You will no doubt need `mount`, `ifconfig`, and probably several others to get the system up and running. Remember that every shell command line utility is a separate program, with its own executable file.

The idea behind BusyBox is brilliantly simple. Rather than have every utility be a separate program with its attendant overhead, why not simply write one program that implements *all* the utilities? Well, perhaps not all, but a very large number of the most common utilities. Most utilities require the same set of "helper" functionality, such as parsing command lines and converting ASCII to binary. Rather than duplicating these functions in hundreds of files, BusyBox implements them exactly once.

The BusyBox project began in 1996, with the goal of putting a complete Linux system on a single floppy disk that could serve as a *rescue disk* or an installer for the Debian Linux distribution. A rescue disk is used to repair a Linux system that has become unbootable. This means the rescue disk must be able to boot the system and mount the hard disk file systems, and it must provide sufficient command line tools to bring the hard disk root file system back to a bootable state.

Subsequently, embedded developers figured out that this was an obvious way to reduce the Linux footprint in resource-constrained embedded environments. So the project grew well beyond its Debian roots, and today BusyBox is a part of almost every commercial embedded Linux offering, and is found in a wide array of Linux-based products, including numerous routers, and media players.

BusyBox calls itself the "Swiss army knife" of embedded Linux because, like the knife, it is an all-purpose tool. Technically, the developers refer to it as a "multi-call binary," meaning that the program is invoked in multiple ways to execute different commands. This is done with symbolic links named for the various utilities. These links then all point to the BusyBox executable.

## Configuring and Installing BusyBox

The version of BusyBox running on my BBB is 1.20.2. Go to the website in the "Resources" section and download that version, or any other one that matches your board or strikes your fancy. Oddly, even though the BBB incorporates BusyBox, it appears there are only two links to it. This may be because we are running a Debian distro, and Debian itself does not use BusyBox.

BusyBox is highly modular and configurable. While it is capable of implementing over 300 hundred shell commands, by no means are you required to have all of them in your system. The configuration process lets you choose exactly which commands will be included in your system. Table 14.1 is the full list of commands available in recent releases of BusyBox.

**Table 14.1: BusyBox commands**

| [ | dirname | hwclock | md5sum | resize | tac |
|---|---|---|---|---|---|
| [[ | dmesg | id | mdev | rm | tail |
| acpid | dnsd | ifconfig | mesg | rmdir | tar |
| addgroup | dnsdomainname | ifdown | microcom | rmmod | taskset |
| adduser | dos2unix | ifenslave | mkdir | route | tcpsvd |
| adjtimex | dpkg | ifplugd | mkdosfs | rpm | tee |
| ar | du | ifup | mkfifo | rpm2cpio | telnet |
| arp | dumpkmap | inetd | mkfs.minix | rtcwake | telnetd |
| arping | dumpleases | init | mkfs.vfat | run-parts | test |
| ash | echo | inotifyd | mknod | runlevel | tftp |
| awk | ed | insmod | mkpasswd | runsv | tftpd |
| basename | egrep | install | mkswap | runsvdir | time |
| beep | eject | ionice | mktemp | rx | timeout |
| blkid | env | ip | modprobe | script | top |
| brctl | envdir | ipaddr | more | scriptreplay | touch |
| bunzip2 | envuidgid | ipcalc | mount | sed | tr |
| bzcat | expand | ipcrm | mountpoint | sendmail | traceroute |
| bzip2 | expr | ipcs | mt | seq | true |
| cal | fakeidentd | iplink | mv | setarch | tty |
| cat | false | iproute | nameif | setconsole | ttysize |
| catv | fbs plash | iprule | nc | setfont | udhcpc |
| chat | fdflush | iptunnel | netstat | setkeycodes | udhcpd |
| chattr | fdformat | kbd_mode | nice | setlogcons | udpsvd |
| chgrp | fdisk | kill | nmeter | setsid | umount |
| chmod | fgrep | killall | nohup | setuidgid | uname |
| chown | find | killall5 | nslookup | sh | uncompress |
| chpasswd | findfs | klogd | od | sha1sum | unexpand |
| chpst | flash_lock | last | openvt | sha256sum | uniq |
| chroot | flash_unlock | length | passwd | sha512sum | unix2dos |
| chrt | fold | less | patch | showkey | unlzma |
| chvt | free | linux32 | pgrep | slattach | unlzop |
| cksum | freeramdisk | linux64 | pidof | sleep | unzip |
| clear | fsck | linuxrc | ping | softlimit | uptime |
| cmp | fsck.minix | ln | ping6 | sort | usleep |
| comm | fsync | loadfont | pipe_progress | split | uudecode |
| cp | ftpd | loadkmap | pivot_root | start-stop-daemon | uuencode |
| cpio | ftpget | logger | pkill | stat | vconfig |
| crond | ftpput | login | popmaildir | strings | vi |
| crontab | fuser | logname | printenv | sty | vlock |
| cryptpw | getopt | logread | printf | su | volname |

*(Continued)*

**Table 14.1: (Continued)**

| | | | | | |
|---|---|---|---|---|---|
| cut | getty | losetup | ps | sulogin | watch |
| date | grep | lpd | pscan | sum | watchdog |
| dc | gunzip | lpq | pwd | sv | wc |
| dd | gzip | lpr | raidautorun | svlogd | wget |
| deallocvt | hd | ls | rdate | swapoff | which |
| delgroup | hdparm | lsattr | rdev | swapon | who |
| deluser | head | lsmod | readlink | switch_root | whoami |
| depmod | hexdump | lzmacat | readprofile | sync | xargs |
| devmem | hostid | lzop | realpath | sysctl | yes |
| df | hostname | lzopcat | reformime | syslogd | zcat |
| dhcprelay | httpd | makemime | renice | | zcip |
| diff | hush | man | reset | | |

BusyBox supports the `xconfig` and `menuconfig` make targets for configuration just like the kernel. Fig. 14.1 shows the `xconfig` menu, while Fig. 14.2 shows the top top-level `meuconfig` menu. Also, like the kernel, there is a `make help` target that lists the other make targets. Oddly, `xconfig` does not show in the list.

There are only a few `*_defconfig` targets for creating default configurations, and sadly none of them match the BeagleBone Black. I have not been able to find a default configuration for the BBB, so we are kind of left hanging. Let us see what happens if we just create a new one.

## *BusyBox Settings*

While the xconfig navigation panel shows a number of option categories, there are really only two: options to configure, build, and install BusyBox, and options to select the desired applets. Under `Build Options` you have the choice of building BusyBox either with statically linked libraries, or shared libraries. A statically linked BusyBox is a much bigger executable, but then you do not need the shared libraries. As an example, the BusyBox I built with shared libraries is 743 kb. The statically linked version is 1.5 MB. But the `/lib` directory, where the shared libraries are kept, adds up to 17 MB. So the question becomes, do you need those shared libraries for anything else?

Build Options is also where you can set the cross-compiler prefix. Set it to `arm-linux-`.

Under Installation Options, you have a choice of how the symbolic links are created. They can be soft links, hard links, or script "wrappers." The accompanying box describes the distinction between hard and soft links. Since hard links to the same data all share the same inode, that can be a useful option on systems with a limited number of inodes. Soft links are the default, and usually the best way to go.

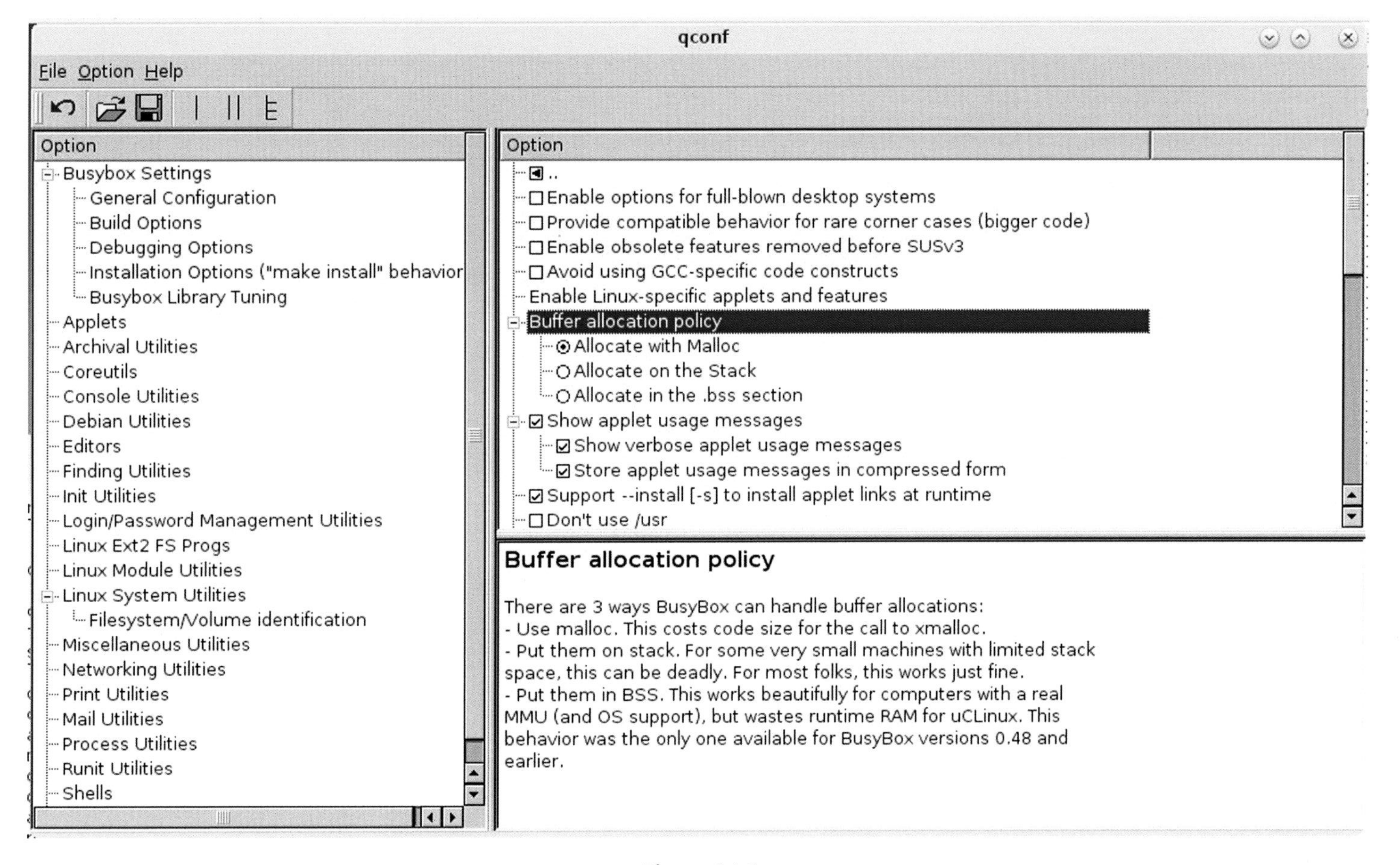


**Figure 14.1**
BusyBox xconfig menu.

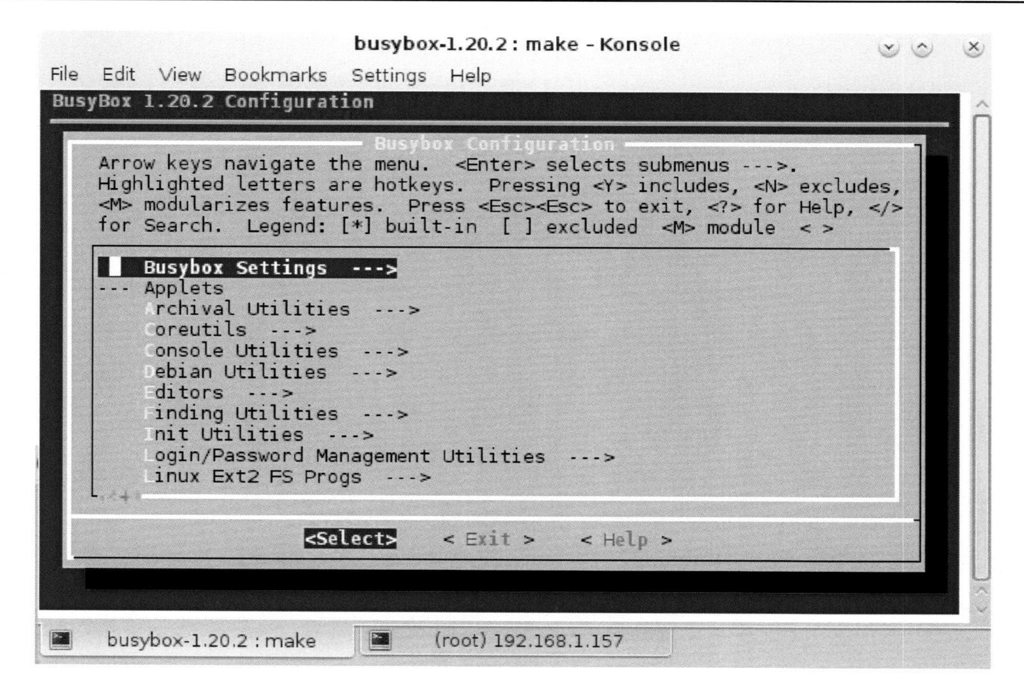


**Figure 14.2**
BusyBox menuconfig top menu.

The Installation Prefix is the root of the file system into which you are installing BusyBox. The default is `./_install`. You can specify it here, or on the command line for the `make install` step by adding `CONFIG_PREFIX = /home//<your_home_directory>/<your_target_fs>`.

## *Applets*

The 300-odd BusyBox applets are divided among several categories. These categories correspond closely to the BusyBox directory tree. In many cases, the BusyBox version of a utility omits some of the more obscure options. You might think of these as "lite" versions. Nevertheless, the options that are available will suffice for just about anything you need in your embedded system.

In some cases, you can select the options you want for a command. For example, select the Coreutils category and scroll down to `ls`. You will find several selectable `ls` options. Presumably, the reason for doing this is to reduce the code size if that is important. It would be nice if help gave some indication of how much space could be saved with each option.

Browse through the categories to get a feel for what is there. You will find that there is an awful lot of stuff enabled that you probably do not need, like the `vi` editor, for example.

And do you really need `tar` and `gzip` on an embedded device? Take a look at the Shells category. The default shell is `ash`, claimed to be the most complete and "pedantically correct" shell in the BusyBox distribution. Two other shells, `hush` and `msh`, offer less functionality in exchange for a smaller footprint.

### Building and Installing

BusyBox uses the normal pattern of `make` and `make install` commands to build and install the package. Following make, you will find several new files in the top-level BusyBox directory, all starting with "busybox." One of them is the actual executable, busybox. Next, execute `make install`, which will copy the busybox executable and all of the links to the destination you specified during configuration. That is it—BusyBox is ready to go.

#### Hard vs. Soft links

Back in Chapter 3 we briefly discussed the idea of "links," entries in the file system that just point to other files, without considering the distinction between *hard* links vs. *soft* or *symbolic* links.

Try this. In a shell window on your workstation go to `~/<your_target_fs>/home/samples/src/led`. Execute these commands:

```
ln led.c led2.c
ln -s led.c led3.c
ls -l
stat led.c
stat led2.c
stat led3.c
```

What do you notice? In particular, note the reference counts, the number just after the permissions field, and the file sizes. `led3.c` is identified as a link that points to `led.c`. Its length is 5, the size of the file *name* it is pointing to. By contrast,`led2.c` and `led.c` have the same length and a reference count of 2, whereas everything else has a reference count of 1. The `stat` commands tell us what inode represents each of the files. The thing to notice is that `led.c` and `led2.c` share the same inode, but `led3.c` has a different inode.

Now do this:

```
rm led.c
ls -l
```

On my CentOS 7, the entry for `led3.c` is highlighted to show that the file it points to does not exist. The reference count for `led2.c` is now 1.

Finally, rename `led2.c` as `led.c`:

```
mv led2.c led.c
ls -l
```

(*Continued*)

**Hard vs. Soft links (Continued)**

The `led3.c` entry is once again pointing at a valid entry.

The point of this exercise is that hard links are alternate names, or aliases if you will, for the same *data*. They all point to the same data blocks on the disk. That data cannot be deleted until the reference count goes to zero. By contrast, a soft or symbolic link points, not at data, but at a file *name*, which may or may not exist.

## Using BusyBox

The BusyBox executable is usually installed in the `/bin` directory, where most user-level command line utilities reside. Then, for example, in `/bin` we create the following symbolic link:

```
ln –s busybox ls
```

Now when you invoke `ls` from a shell, what really gets executed is BusyBox. The program figures out what it is supposed to do by looking at the first argument to `main()`, which is the name that invoked it, i.e., `ls`. A similar link is created for every command that BusyBox implements.

You can also invoke BusyBox directly, passing a command name and its arguments as a command line, like this:

```
busybox ls –l
```

Executing `busybox –help` or just `busybox` by itself yields a help message including a list of the commands built into the current instance of BusyBox.

## User Space Initialization

We now have a useful thermostat application. The next thing we need to do is arrange for that application to start automatically when the system is powered up. Also, we probably do not need a console shell in the final product.

Let us start by examining the initialization process from the moment power is turned on until the Linux kernel starts the first user space process. There are basically four steps in this process.

### Stage 1 Boot Loader

The AM335x Sitara has an extensive built-in boot ROM with several boot options. When power is applied and the reset pin is released, execution begins at the Reset exception vector at 0x20000. This is a jump instruction into what TI calls the Public ROM Code. This code does some very low-level initialization of things like clocks and the watchdog timer. It then builds a list of bootable devices based on the state of the SYSBOOT pins. For each value of the SYSBOOT pins, the boot loader builds a list of four devices to test for a bootable image. Each device is checked in turn, to see if there is a bootable image. In the case of the BBB, there are two options determined by pushbutton S2. The default is:

1. *eMMC*. If the eMMC does not contain a bootable image, then...
2. *microSD* card. If there is no microSD card, or it does not have a bootable image...
3. *debug serial port*. If there is nothing there...
4. *USB port*.

If S2 is pressed when power is applied, the order is:

1. *SPIO*. This will always fail on the BBB, so...
2. *microSD* card. If there is no microSD card, or it does not have a bootable image...
3. *USB port*.
4. *debug serial port*. If there is nothing there...

Oddly enough, the mere presence of a microSD card causes my board to boot from the card. The BeagleBoard folks do not support booting over USB or serial ports. You have to go to TI for that.

### Stage 2 MLO

The Sitara SoC also contains 128 kb of on-board RAM, of which 109 kb is available to load a boot image. This is where the second stage boot loader is loaded. On the BBB the second stage loader is an image file named MLO. Its job is to set up the chip selects for RAM and various off-chip peripherals, and then load the third stage loader, u-boot.

### U-Boot

The u-boot loader continues the initialization process by initializing the peripherals it needs, typically the serial port, network, and may be USB. Then, if the auto boot sequence is not interrupted, it will usually copy the kernel from Flash to RAM and start it executing.

### Linux Kernel

If the kernel is a compressed image, it uncompresses itself into the appropriate location in RAM. Then it initializes all of the kernel subsystems, such as memory management and drivers for all of the peripheral devices in the system. To some extent, this may duplicate the initialization done by U-boot, but of course the kernel has no idea what U-boot did, and so must initialize the devices to suit its own requirements. During the initialization process the kernel spews out a large number of messages describing what it is doing.

Next it mounts the root file system. Up to this point, the kernel has been running in kernel space. Finally it starts the init process, which makes the transition to user space.

### *Systemd*

The last thing the kernel boot does is start a process with PID 1. This then becomes the ultimate parent of every other process in the system. Not too long ago this was a fairly straightforward process involving an executable, `/sbin/init`, a configuration file, `inittab`, and a handful of scripts. This was the System V initialization mechanism.

Now there is a "new improved" initialization mechanism called *systemd* encompassing something like 900 files. By now most major Linux distros have adopted systemd, but it has been a highly contentious issue since its introduction in 2011. Proponents insist that the System V init is too slow, because it starts processes serially, one at a time, whereas systemd parallelizes much of the activity. Opponents say "if it ain't broke, don't fix it."

systemd manages and acts on objects called *units*. There are many types of units, but the most common type is a *service*, represented by a file that ends in `.service`. Have a look in `/lib/systemd/system` of your target file system. You will see lots of `*.service` files. As an example, open up `bonescript.service`. The `WorkingDirectory` generally contains support files for the executable, which is identified by `ExecStart`.

You manage systemd through the `systemctl` command. The most common commands that systemctl understands are:

| | |
|---|---|
| `start <unit>` | Start the specified unit(s) |
| `stop <unit>` | Stop the specified unit(s) |
| `restart <unit>` | Restart the specified unit(s). If not running they will be started |
| `reload <unit>` | Ask specified unit(s) to reload their configuration files |
| `status <unit>` | Display the status of the specified unit(s) |
| `enable <unit>` | Create symlinks to allow unit(s) to be automatically started at boot |
| `disable <unit>` | Remove the symlinks that cause the unit(s) to be started at boot |
| `daemon-reload` | Reloads the systemd manager's configuration. Run this any time you make a change to a systemd file. |

As a first experiment with `systemctl`, just enter the command by itself with no arguments. You will get a list of all the services that are available on the BBB. `systemctl` pipes its output to `more` (or `less`, I don't know) so it is presented one page at a time. There are a lot of services available.

Our objective then is to get the thermostat to start executing when the device boots. We will do that by creating a thermostat service. Remember that we have two simple scripts in /home/samples/src, leds-off.sh to turn off the triggers for the LEDs, and adc-on.sh to enable the A/D converter. Create the script file thermo.sh, as shown in Listing 14.1 in the src/network directory. This script invokes the two scripts just mentioned, and then starts thermostat_t. This is the executable for our service.

```
#!/bin/bash
#

/home/samples/src/leds-off.sh
/home/samples/src/adc-on.sh
/home/samples/src/network/thermostat_t
```

**Listing 14.1**
thermo.sh.

Make a copy of bonescript.service, calling it thermostat.service. Listing 14.2 shows the final version of thermostat.service with the changes and additions underlined.

```
[Unit]
Description=Networked thermostat server

[Service]
WorkingDirectory=/home/samples/src/ network
ExecStart=/home/samples/src/network/thermo.sh
StandardOutput=tty
SyslogIdentifier=thermo

[Install]
WantedBy=multi-user.target
```

**Listing 14.2**
thermostat.service.

Save the file in /lib/systemd/system.

Now execute these systemctl commands:

```
systemctl daemon-reload
systemctl enable thermostat.service
systemctl start thermostat.service
```

systemctl daemon-reload causes systemd to pick up the fact that you added the thermostat.service. The systemctl start command causes the thermostat to start running. The systemctl enable command creates a link in /etc/systemd/system/multi-user.target pointing to /lib/

systemd/system/thermostat.service. This is what causes the thermostat to start at boot up. Reboot the board and the thermostat should start running as soon as the kernel boots up.

To restore the command shell operation, remove the link /etc/systemd/system/multi-user.wants/thermostat.service that points to /lib/systemd/system/thermostat.service.

### User Space Initialization, the "Old Way"

As of version 7.4, systemd was considered an "experimental" feature of Debian. The System V initialization code is still there. Take a look at uEnvnet.txt. Near the top, the variable systemd is defined to redirect init to systemd. That variable is used at the end of the netargs definition. To restore the System V initialization, simply remove the variable from the netargs definition.

The init executable is typically /sbin/init, although there are several alternative locations that the kernel will search. init gets its instructions from the file /etc/inittab. Take a look at that file. The comments at the top give a pretty good idea of what is going on. Each entry has four fields separated by colons:

```
<id>:<runlevel>:<action>:process
```

| | |
|---|---|
| id | tty connected to the process. NULL if the process does not use a console. |
| runlevel | one or more of the seven run levels that Linux can start up in. This line will be executed for the specified run levels |
| action | one of eight ways init will treat the process |
| process | program to run including arguments if any |

The allowable actions are:

| | |
|---|---|
| once | execute the process once |
| wait | execute the process once. Init waits until the process terminates |
| askfirst | ask the user if this process should be run |
| sysinit | these processes are executed once before anything else executes |
| respawn | restart the process whenever it terminates |
| restart | like respawn |
| shutdown | execute these processes when the system is shutting down |
| ctrlaltdel | execute this when init receives a SIGINT signal, meaning the operator typed CTRL-ALT-DEL. |

Note that the sysinit action in this case is to execute the rcS script.

Probably the easiest way to start up our application is from within inittab. Find the line near the bottom that reads:

```
T0:23:respawn:/sbin/getty -L ttyO0 115200 vt102
```

Duplicate that line and change it to:

```
th:23:respawn:/home/samples/src/network/thermo.sh
```

We do not need a console terminal, so comment out the existing line that starts with `T0:`. Save `inittab`. Remove `${systemd}` from the `netargs` definition in `uEnv.txt` and replace the file in the boot partition.

Reset the target and let it boot up Linux. You should see the thermostat start up.

Note, by the way, that another approach to starting the application is to simply replace `/sbin/init`. Make it a symbolic link that points to the application executable. This may be fine for simple applications, but there is a lot of functionality in `init` that might be useful in the final product.

The other thing you might want to consider at this point is building a minimal root file system based largely on BusyBox. The resources section lists a couple of blog posts about that.

In the next chapter we will take a closer look at U-boot, and get our "product" ready to "ship."

## Resources

www.busybox.net—This is the official BusyBox website.

access.redhat.com/documentation/en-US/Red_Hat_Enterprise_Linux/7/html/System_Administrators_Guide/sect-Managing_Services_with_systemd-Unit_Files.html—Pretty thorough documentation of systemd.

There are a number of tutorials on systemd. Just google "systemd tutorial." Here are a couple I found useful:

www.linux.com/learn/understanding-and-using-systemd

www.digitalocean.com/community/tutorials/systemd-essentials-working-with-services-units-and-the-journal

Here are a couple of blog posts on building minimal BeagleBone Black systems from scratch.

gist.github.com/vsergeev/2391575

www.bootembedded.com/embedded-linux/building-embedded-linux-scratch-beaglebone-black

CHAPTER 15

# U-boot boot loader and getting ready to ship

## Chapter Outline



*The box said "Requires Windows 95 or better." So, I installed Linux.*

## U-Boot

As we saw in the previous chapter, a boot loader is a program that does some initialization in preparation for loading and running the operating system and its supporting infrastructure. In a sense, the boot loader is like the BIOS (Basic Input/Output System) in a desktop PC or server. The principal difference is that a boot loader executes once when the system powers up and then goes away. The BIOS hangs around to provide low-level I/O services.

### Background

Desktop Linux systems have a boot loader in addition to the BIOS. These days it is usually GRUB, the GRand Unified Bootloader, or GRUB2. But because the BIOS does all the heavy lifting of initializing the hardware, GRUB itself does little more than load and start the operating system. If you are building an embedded system based on conventional x86 PC hardware, GRUB is probably the way to go.

On the other hand, our BBB target board uses a very popular, very capable, Open Source, cross-platform boot loader to get things started. We briefly saw some of the features of u-boot back in Chapter 6, The hardware, when we brought the target board up.

Linux for Embedded and Real-time Applications.
DOI: http://dx.doi.org/10.1016/B978-0-12-811277-9.00015-8

U-boot began as a PowerPC boot loader named 8xxROM written by Magnus Damm. Wolfgang Denk subsequently moved the project to Source Forge, and renamed it PPCBoot. The source code was briefly forked into a product called ARMBoot by Sysgo GmbH. The name was changed to u-boot when the ARMBoot fork was merged back into the PPCBoot tree. Today, u-boot supports roughly a dozen architectures, and over 1000 different boards.

The development of u-boot is closely tied to Linux, with which it shares some header files. Some of the source code originated in the kernel source tree.

U-boot supports an extensive command set that not only facilitates booting, but also manages flash memory, downloads files over the network, and more. Appendix I details the command set. The command set is augmented by environment variables and a scripting language.

### *Installing and Configuring U-Boot*

The current version of u-boot is available from the following git repository:

```
git clone git://git.denx.de/u-boot.git
```

There is a very extensive `README` file in the top-level directory. There are additional `README` files in the `doc/` directory that primarily describe features of specific boards.

There is also a patch to the mainline code for the BeagleBoard. Point your web browser to:

https://rcn-ee.com/repos/git/u-boot-patches/

and select the subfolder corresponding to the u-boot version you got from git. Download `0001-am335x_evm-uEnv.txt-bootz-n-fixes.patch` to your u-boot directory. Then execute:

```
patch –p1 < 0001-am335x_evm-uEnv.txt-bootz-n-fixes.patch
```

In my case, two of the eight files failed their patches, but were pretty easy to patch by hand.

Current u-boot implementations support the same Kconfig configuration mechanism as the Linux kernel and BusyBox. Again, we have to start with a default configuration. The u-boot version I downloaded has `_defconfigs` for over a thousand boards. In our case it is:

```
make am335x_boneblack_defconfig
```

Follow that with:

```
make xconfig
```

Fig. 15.1 is the top-level xconfig menu for u-boot. This shows that the ARM (Advanced RISC Machines) architecture is selected along with the TI OMAP2 + variant. There is probably not a whole lot that you would want to change, other than which commands get included.

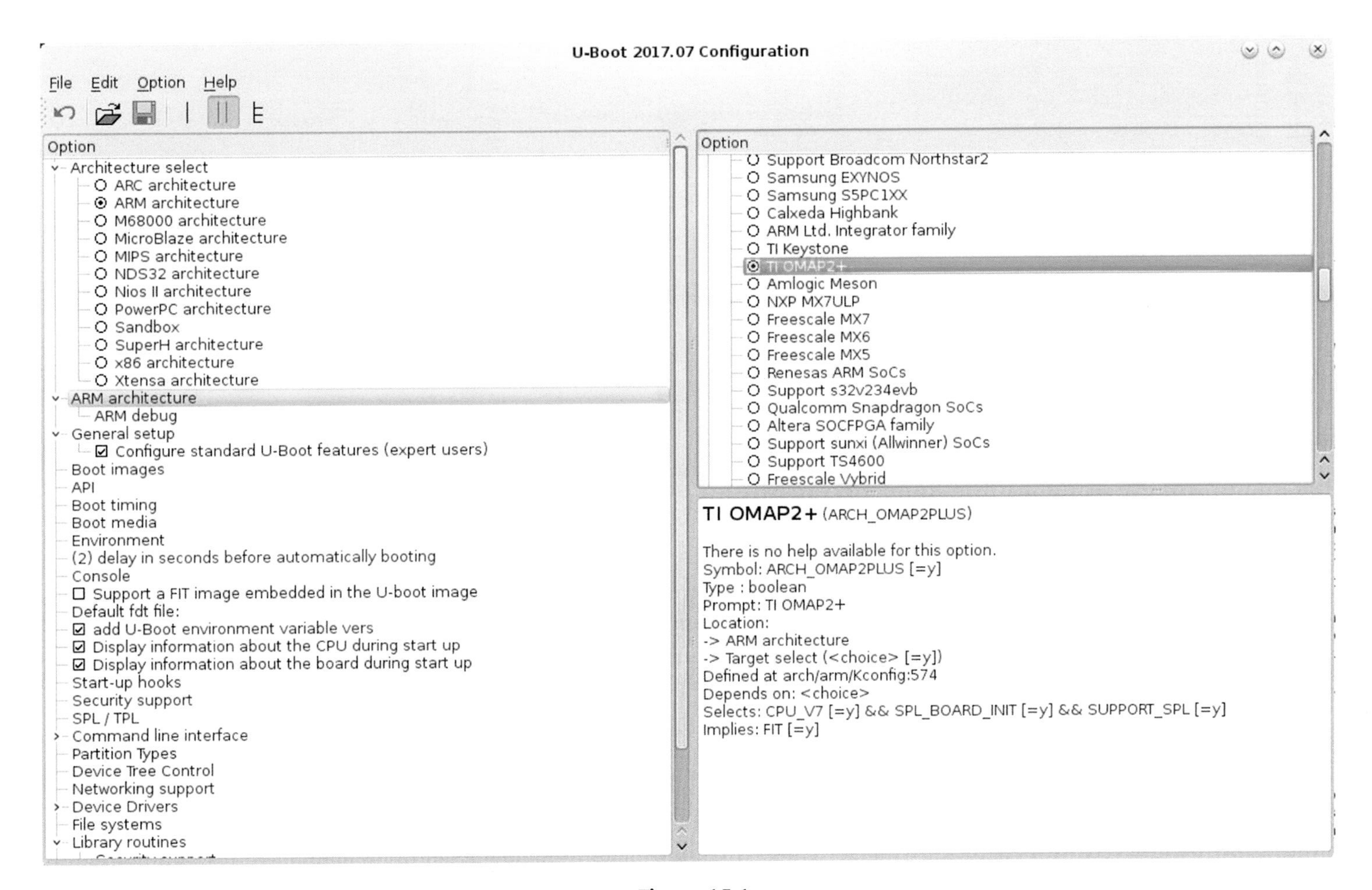


**Figure 15.1**
U-boot make xconfig.

You will note that there is no CROSS_COMPILE prefix in the xconfig menu, so you will have to either edit the Makefile, or enter it in the make command. Also, I have not found where the ARCH variable is defined in the Makefile, so you have to specify that on the make command line. Your make command then looks like this:

```
make CROSS_COMPILE=arm-linux- ARCH=arm
```

I ran into a couple of problems. My first run of make yielded this error message:

```
bad value (armv5) for -march= switch
```

Google did not turn up much useful for this one, at least not with respect to u-boot. Poking around in the Makefiles, I found these two lines in arch/arm/Makefile:

```
arch-$(CONFIG_CPU_V7)          =$(call cc-option, -march=armv7-a, \
                                        $(call cc-option, -march=armv7,
-march=armv5))
```

Indeed, CONFIG_CPU_V7 is set, so these lines get added to the gcc command line. But why the –march=armv5? As an experiment I removed the ", -march=armv5" from the second line and the make succeeded.

The other issue is you need the package python-devel.

The make step generates several output files:

- MLO—the secondary program loader (SPL) that loads u-boot into RAM
- u-boot—the executable in ELF binary format. It is not clear what this file is for
- u-boot.bin—the raw binary image, suitable for writing to flash
- u-boot-nodtb.bin—appears to be identical to u-boot.bin
- u-boot.img—appears to be u-boot.bin with the 64-byte u-boot header prepended to it
- u-boot.map—map file output by the linker
- u-boot.cfg—the .config file expressed as a C header file
- u-boot.sym—symbol table
- u-boot.srec—the executable in Motorola S record format

The make step also builds several tools in the tools/ directory.

### Testing a New U-Boot

You will want to test your new u-boot initially in RAM before burning it to flash. Copy u-boot.bin to /var/lib/tftpboot. Boot the BBB into u-boot and execute:

```
setenv ipaddr 192.168.1.50
setenv serverip 192.168.1.2
```

```
tftp 80800000 u-boot.bin
go 80800000
```

We have to set the `ipaddr` and `serverip` because those normally do not get set until u-boot reads `uEnv.txt`. The `go` command transfers control to the specified memory address. The new boot loader should start executing and boot the kernel, unless you interrupt the autoboot.

You can also use a JTAG (Joint Test Action Group) hardware debugger if you happen to have one. This is useful if you have made substantial changes and really need to test the new u-boot, because it is probably the only way to use breakpoints on boot loader code.

### *The "Sandbox"*

If your modifications to u-boot are architecture independent, there is a way to do your testing on your workstation. U-boot provides a target "board," called the "sandbox," in the spirit of the simulation environment we looked at in Chapter 8, Debugging embedded software. The sandbox runs u-boot as a user space application that allows you to use Gnu DeBugger (GDB) to debug your modifications. To build the sandbox version you will probably need to install the package `openssl-devel`.

By default, the sandbox requires something called `libsdl`, where SDL stands for Simple DirectMedia Layer. The u-boot documentation implies that this is necessary for display and keyboard support. The package is supposedly called `libsdl1.2-dev`, which I could not find in the CentOS repos. You can build the sandbox without SDL support by adding `NO_SDL=1` to the make command.

That is not the whole story, though. You have to edit `include/configs/sandbox.h`. Down around line 104 are these there lines:

```
#ifndef SANDBOX_NO_SDL
#define CONFIG_SANDBOX_SDL
#endif
```

Comment those lines out. Now execute:

```
make NO_SDL=1
```

The build appears to succeed, but I got the following messages at the end:

```
./scripts/dtc-version.sh: line 17: dtc: command not found
./scripts/dtc-version.sh: line 18: dtc: command not found
*** Your dtc is too old, please upgrade to dtc 1.4 or newer
make: *** [checkdtc] Error 1
```

Dtc refers to the Device Tree Compiler, which we will get to shortly. Nonetheless, the u-boot executable is there, and you can run it from a command shell. The device tree compiler is part of the Linux kernel source tree and again, there does not seem to be a corresponding package in the CentOS repos. A simple hack to solve this one is to copy `linux/scripts/dtc/dtc` to somewhere in your path such as `/usr/local/bin`.

### Device Trees

One of the biggest problems with porting an operating system such as Linux to a new platform is describing the hardware. That is because the hardware description is scattered over perhaps several dozen or so device drivers, the kernel, and the boot loader, just to name a few. The ultimate result is that these various pieces of software become unique to each platform, the number of configuration options grows, and every board requires a unique kernel image.

There have been a number of approaches to addressing this problem. The notion of a "board support package" or BSP attempts to gather all of the hardware-dependent code in a few files in one place. It could be argued that the entire `arch/` subtree of the Linux kernel source tree is a gigantic board support package.

Take a look at the `arch/arm/` subtree of the kernel. In there you will find a large number of directories of the form `mach-*` and `plat-*`, presumably short for "machine" and "platform," respectively. Most of the files in these directories provide configuration information for a specific implementation of the ARM architecture. And of course, each implementation describes its configuration differently.

Would not it be nice to have a single language that could be used to unambiguously describe the hardware of a computer system? That is the premise, and promise, of device trees.

The peripheral devices in a system can be characterized along a number of dimensions. There are, for example, character vs block devices. There are memory mapped devices, and those that connect through an external bus such as I2C or USB. Then there are *platform* devices and *discoverable* devices.

Discoverable devices are those that live on external busses, such as PCI and USB, that can tell the system what they are and how they are configured. That is, they can be "discovered" by the kernel. Having identified a device, it is a fairly simple matter to load the corresponding driver, which then interrogates the device to determine its precise configuration.

Platform devices, on the other hand, lack any mechanism to identify themselves. System on Chip (SoC) implementations, such as the Sitara, are rife with these platform devices—

system clocks, interrupt controllers, GPIO, serial ports, to name a few. The device tree mechanism is particularly useful for managing platform devices.

The device tree concept evolved in the PowerPC branch of the kernel, and that is where it seems to be used the most. In fact, it is now a requirement that all PowerPC platforms pass a device tree to the kernel at boot time. The text representation of a device tree is a file with the extension .dts. These `.dts` files are typically found in the kernel source tree at `arch/$ARCH/boot/dts`.

A device tree is a hierarchical data structure that describes the collection of devices and interconnecting busses of a computer system. It is organized as nodes that begin at a root represented by "/," just like the root file system. Every node has a name and consists of "properties" that are name-value pairs. It may also contain "child" nodes.

Listing 15.1 is a sample device tree taken from the devicetree.org website. It does nothing beyond illustrating the structure. Here we have two nodes named `node1` and `node2`. `node1` has two child nodes, and `node2` has one child. Properties are represented by `name = value`. Values can be strings, lists of strings, one or more numeric values enclosed by square brackets, or one or more "cells" enclosed in angle brackets. The value can also be empty if the property conveys a Boolean value by its presence or absence.

```
/{
   node1 {
      a-string-property = "A string";
      a-string-list-property = "first string", "second string";
      a-byte-data-property = [0x01 0x23 0x34 0x56];
      child-node1 {
         first-child-property;
         second-child-property = <1>;
         a-string-property = "Hello, world";
      };
      child-node2 {
      };
   };
   node2 {
      an-empty-property;
      a-cell-property = <1 2 3 4>; /* each number (cell) is a uint32 */
      child-node1 {
      };
   };
};
```

**Listing 15.1**
Sample device tree.

The BBB uses a device tree to describe its hardware. The device tree source files (`*.dts`) are found in `linux/arch/arm/boot/dts`. The one for the BBB is `am335x-boneblack.dts`. This file does not define much, but it includes a couple of other files (note that included device tree files get the extension `.dtsi`). `am33xx.dtsi` defines quite a bit of the system. Note, in particular, that pretty much every node has a *compatible* property. This is what links a node to the device driver that manages it.

The device tree source code is compiled into a more compact form called a *device tree blob* (*.dtb), also known as a *flattened device tree* (fdt). As mentioned above, the device tree compiler is actually part of the kernel source tree. The resulting `am335x-boneblack.dtb` is stored in the same flash partition as `MLO` and `u-boot.img`. The default u-boot environment has a variable, `loadfdt`, that loads the device tree blob into RAM. Then the `bootz` command that transfers control to the kernel has three address arguments: the address of the kernel, the address of the initial RAM disk, and the address of the device tree blob. As the kernel loads device drivers, the drivers interrogate the device tree to determine the specific properties of their respective devices.

The device tree is visible from u-boot using the `fdt` commands. Type `help fdt` to see the commands. Among other things, you can list the fdt, create new nodes and properties, and change property values. Then once the kernel is booted, the device tree is visible at `/proc/device-tree`. This tree is read-only.

## Putting the Application in eMMC Flash

We are almost done. We have configured the kernel and the boot loader to our specific needs. We have set up systemd and/or the init process so that it boots directly into the thermostat application. At the end of the last chapter, we ended up with a file system that booted directly into our networked thermostat application. The final step then is to load that file system into eMMC flash on the target board, so that when the board powers up, the thermostat starts.

The process is actually quite straightforward. You do still have the microSD card with the eMMC flasher image on it, the one we used back in Chapter 6, The hardware, right? Here are the steps then:

- Insert the microSD card into your workstation
- Mount both partitions, BEAGLEBONE and eMMC-Flasher
- Copy any changes from your networked file system to the microSD eMMC-Flasher partition
- Restore the original `uEnv.txt` to the BEAGLEBONE partition so the device boots from eMMC
- If necessary, copy your new `u-boot.img` and kernel `zImage` files to the BEAGLEBONE partition
- Insert the microSD card into your unpowered BBB and power up
- The eMMC is flashed with your "production" file system
- Power down, remove the microSD card, and power up. Your "product" is now running.

There are a couple other options for flashing the eMMC. Robert Nelson, an applications engineer with DigiKey, has done substantial work on the BeagleBone and BBB. Among other things, he has created a collection of scripts for building eMMC flasher images, and then flashing them to the eMMC. This might be a useful approach if, for example, you wanted to build your own minimal root file system. He has another project to do a net install. The resources section has the links for both.

OK, the thermostat application is now burned into flash, and we are ready to ship. In the next chapter we will turn our attention to source code control.

## Resources

www.denx.de—This is the website for Wolfgang Denk, the principal developer of u-boot. There is a good user's manual here, a wiki, and of course, the source code at:

ftp://ftp.denx.de/pub/u-boot/—Versions of u-boot going back to 2002 are available here.

wiki.beyondlogic.org/index.php?title = BeagleBoneBlack_Upgrading_uBoot—This page alerted me that there is a BeagleBone Black patch for u-boot.

devicetree.org—Home page for the device tree project. Among other things, this site documents bindings that are not covered by the Linux kernel or the ePAPR (see next reference). There is also a good tutorial on device tree usage.

*Standard for Embedded Power Architecture™ Platform Requirements ePAPR*—Available at: www.power.org/resources/downloads/Power_ePAPR_APPROVED_v1.0.pdf. Probably the most complete definition of the device tree.

github.com/RobertCNelson/omap-image-builder—Collection of scripts for building eMMC flasher images and then flashing them.

github.com/RobertCNelson/netinstall—Collection of scripts for installing ARM images over the network.

CHAPTER 16

# Source code control — GIT

## Chapter Outline



*And then realize that nothing is perfect. GIT is just closer to perfect than any other SCM out there.*

**Linus Torvalds**

## Background

Source Code Control, also known as Revision Control, refers to the practice of storing the source code files and other artifacts of a project, such as documentation, in a common repository so that multiple developers can work on the project simultaneously without interfering with each other. The source code control (SCC) software maintains a record of changes and supports multiple versions of a project simultaneously. It also allows you to go back and recreate previous versions if necessary.

There are a number of SCC packages, both open source and proprietary, in widespread use. Most of these follow a centralized client/server model where the repositories reside on a central server as illustrated in figure 16.1. Developers "check out" individual project files, work on them and subsequently check them back in. Most SCCs allow multiple developers to edit the same file at the same time. The first developer to check in changes to the central repository always succeeds. Many systems provide facilities to automatically *merge* subsequent changes provided they don't conflict.

Linux for Embedded and Real-time Applications.
DOI: http://dx.doi.org/10.1016/B978-0-12-811277-9.00016-X

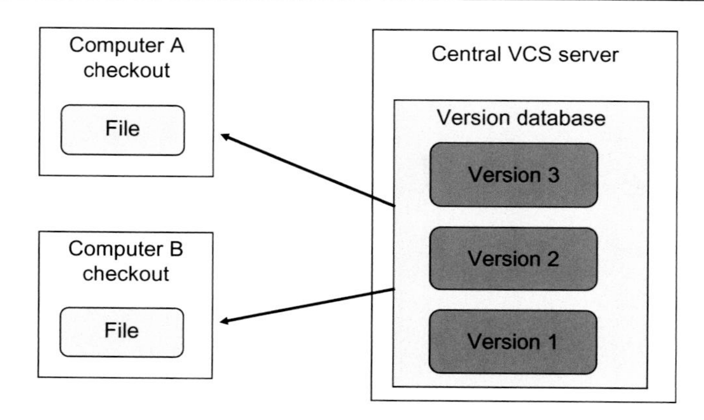


**Figure 16.1**
Centralized version control.

Some of the better known SCC systems include:

- CVS
- BitKeeper
- Rational Clear Case
- Mercurial
- Perforce
- Subversion
- Visual Source Safe

Until 2002, the kernel development community didn't use a version control system (VCS). Changes were informally passed around as patches and archive files. In 2002 the community began using BitKeeper, a proprietary distributed VCS. Subsequently, friction between the kernel developers and the BitKeeper people resulted in the former losing their free usage of the package.

This prompted the Linux community to develop their own tool based on lessons learned while using BitKeeper. The resulting system features a simple design that runs incredibly fast while satisfying these goals:

- Speed
- Simple design
- Fully distributed
- Support for "non-linear" development
    - Potentially thousands of parallel branches
- Handles large projects such as the kernel efficiently

So where did the name "git" come from? Git is British slang for a stupid or unpleasant person. As Linus explained it, "I'm an egotistical bastard, and I name all my projects after myself. First 'Linux', now 'git'."

## Introducing git

Git is a *distributed* version control system where clients don't just check out snapshots of files, they fully mirror the central repository. This is illustrated in figure 16.2. Each

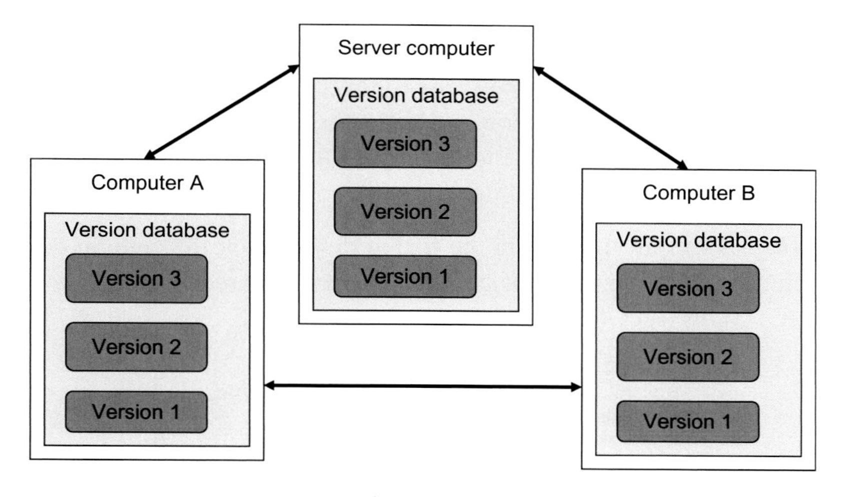


**Figure 16.2**
Distributed version control.

checkout is a full backup. So if the central server dies, it can be restored from any of the clients with little or no loss. Git is installed with the development tools in most modern Linux distributions.

Figure 16.3 shows a typical sequence of git commands. You start by getting a copy of a project repository with the `git clone` command. This creates a complete local copy of the project from the currently active branch of the central repository. The `hello-world` project is hundreds of versions of the Hello World program in different languages.

Then you make changes to the project. Just for practice, `cd` to the `c/` directory and edit `c.c` to add a header comment. `git add` adds the changes you've made to the *index* in preparation for the next *commit*. The index holds a snapshot of the changes that have been added since the last commit. You can invoke `git add` multiple times between commits.

```
$ git clone https://github.com/ leachim6 /hello-world.git
$ cd hello-world
        Edit files
$ git add (files)
$ git commit –m 'Explain what I changed'
$ git format-patch origin/master
$ git push
```

**Figure 16.3**
git command sequence.

`git commit` stores the index back to the local repository along with a comment describing what changed. The `–m` option means a *message* follows. This is good for relatively short log messages. If you leave off the −m option, git opens the environment variable `$EDITOR`, allowing you to create a more detailed log message. The significant point is that git insists you provide a comment every time you do a commit.

A *commit* is a snapshot of all the files in the tree at the moment the commit command is executed. Differences from the previous commit or snapshot are stored in a compressed delta form under `.git/objects` {{{Figure 16.4}}}.

```
$ git init          Create a new empty repository in the
current directory
$ git add *         Add all existing files to the index
$ git commit –m 'initial project version'
$ git push          Send to a remote repository
```

**Figure 16.4**
Creating a new repository.

When you create a new repository, a new subdirectory, `.git/`, shows up in the directory where you created the repository. `.git/` contains everything git needs to know about the project. Try it with the `network/` project.

`.gitignore` is a file you should add to every repository you create. It contains a list of files that don't need to be tracked by git. You don't need to track object files (`*.o`) or backups (`*~`) for instance. `hello-world/` has an example of a `.gitignore` file.

### *File states and life cycle*

From git's perspective, every file in your working directory exists in one of two states: tracked or untracked. Tracked files are the ones git knows about and "tracks". Tracked files may be unmodified, modified, or staged. When you clone a project, all the files are initially tracked and unmodified. You determine the state of project files with the `git status` command.

Try this. Add a `README` file to the `hello-world` project and then execute `git status`. The result will look something like listing 16.1. In this case there's also a modified file that has been

```
[doug@ldougs hello-world]$ git status
# On branch master
# Changes to be committed:
#   (use "git reset HEAD <file>..." to unstage)
#
#       modified:   c.c
#
# Untracked files:
#   (use "git add <file>..." to include in what will be committed)
#
#       README
```

**Listing 16.1**
`git status`.

*staged* (by executing git add on it) but hasn't been committed yet. As the status output says, use `git add<file>` to track the file and include it in the next commit.

Execute `git add README`. `README` is now *staged*, that is, ready to be committed, along with any files that have been modified and added since the last commit.

This brings us to the state diagram of figure 16.5. The primary purpose of this state machine view of git is to make the point that most operations are performed locally, off line. This has two primary advantages:

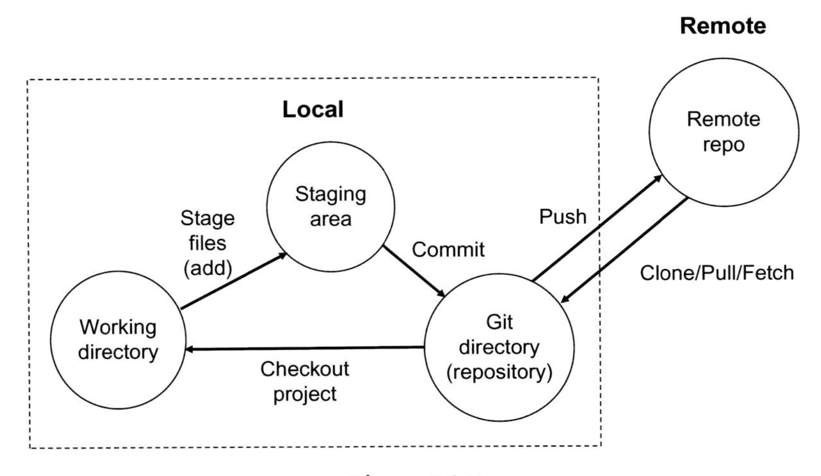


**Figure 16.5**
git state machine view.

- It's faster.
- You don't have to be on line to work.

We've already seen `git clone` and `git push`. `git fetch` retrieves updates from the remote repository. `git pull` does a fetch followed by a merge to integrate the updates into the current branch.

### Branching and merging

Perhaps git's most outstanding feature is its branching model. Other VCSs support branching, but git does so in a particularly elegant way. In fact, some users call the branching model git's "killer feature".

To understand how branching works, we need to step back and take a closer look at how git stores commits. Figure 16.6 shows the contents of a single commit. It consists of a commit

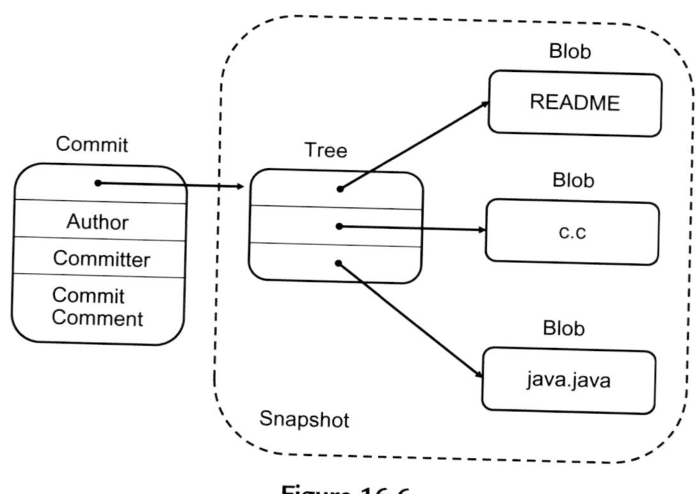


**Figure 16.6**
Representation of a commit.

metadata object that contains the author, committer, and commit comment among other things, plus a pointer to a tree object that in turn points to *blobs* representing individual objects that have been changed in this commit. The tree and blobs constitute the *snapshot* that this commit represents.

Figure 16.7 shows the situation after two more commits. Each subsequent commit object points to its *parent*, the commit from which it was created. This is how we can recreate the complete history of the project. There are two other objects in this diagram. Master is the name of the current *branch*, the one that you're currently working in. Every project starts with a master branch. `HEAD` points to the currently active branch. `HEAD` is a file in `.git/`.

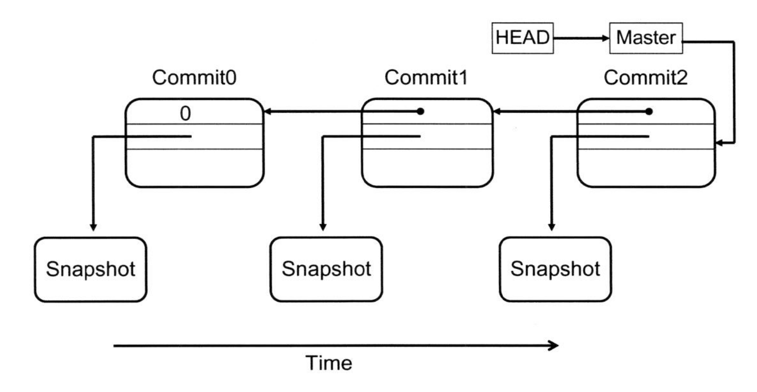


**Figure 16.7**
Multiple commits.

OK, let's create a new branch. Maybe you want to test out some ideas, but don't want to disturb the production code until you've thoroughly tested them. Try it with the `hello-world` project. Execute `git branch testing` followed by `git checkout testing`. The branch command creates a new branch called testing and the checkout command causes `HEAD` to point to the specified branch as illustrated in figure 16.8.

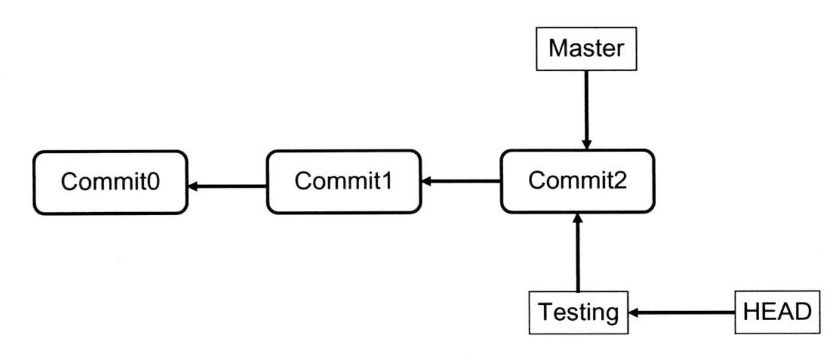


**Figure 16.8**
Adding a branch.

Now make a change to one of the files followed by `git add` and `git commit`. Checkout the master branch and make a change there. That results in the situation of figure 16.9. The branches have diverged. It's important to realize that whenever you checkout a different branch, git restores the state of the working directory to the snapshot that the branch is currently pointing to.

Eventually you'll reach the point where you want to merge the testing branch back into the master production branch. With the master branch checked out, execute `git merge testing`.

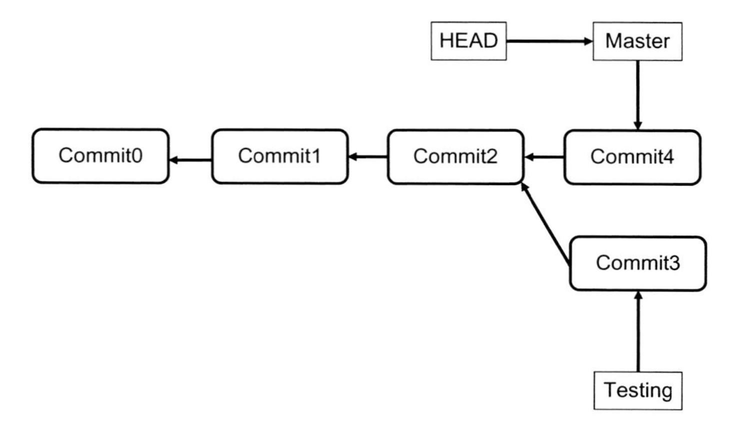


**Figure 16.9**
Diverging branches.

Git performs a 3-way merge among the two divergent branches and a common ancestor, creating a new commit object. See figure 16.10.

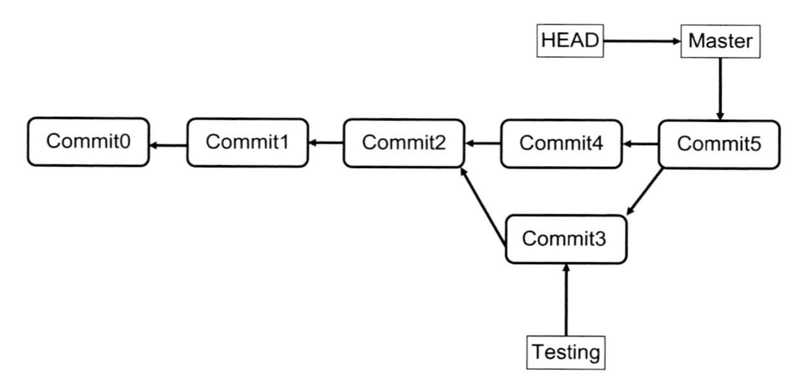


**Figure 16.10**
Merging branches.

At this point you probably no longer need the testing branch. You can delete it with `git merge –d testing`.

## *Configuring git*

When you did your first commit, you may have noticed that git suggested a name and an email address that it pulled from your system information. User name and information are rather important pieces of data because other developers may need to contact you about your commits to the project. So it's important to configure your git package with accurate basic information.

Git stores configuration information in three places:

- `/etc/gitconfig` – System-wide parameters applicable to all users. Use `git config –system` to read or write these values.
- `~/.gitconfig` – Parameters specific to each user. Use `git config –global` to access these.
- `.git/config` – Project-specific parameters found in the repository you're currently working with.

Each level overwrites values in the previous level, so values in `.git/config` take precedence over those in `/etc/gitconfig`, for instance. These files are of course plain ASCII text so you can set the values by manually editing the file and inserting the correct syntax, but it's generally easier to run the `git config` command.

To get started, set your user information:

```
git config –global user.name "Homer Simpson"
git config –global user.email homer@thesimpsons.com
```

These commands hint at the basic structure of the configuration files. All configuration variables are specified as `section.variable`. In the configuration file, the section name is enclosed in square brackets. The square brackets may also include a subsection name enclosed in quotes. That syntax becomes `section.subsection.variable`.

To get a better feel for how all of this hangs together, execute `git config –list`, which lists all of the currently defined configuration variables. Then open `hello-world/.git/config` with an editor to see where those values came from.

Git provides a help command, `git help`, which lists all of the commands. If you then execute `git help<command>`, you'll get the man page for `<command>`. Execute `git help config` and you'll eventually come to a list of configuration variables that git recognizes.

## Graphical git

It should perhaps come as no surprise that Eclipse supports git. Fire up Eclipse and change to the Git perspective. The Project Explorer view has been replaced by Git Repositories view. The menu bar of that view has icons to:

- Add a local repository to the view
- Clone a repository to the view
- Create a new git repository in the view

Since we already have a repository for `hello-world`, let's add that to Eclipse. Click the Add icon to bring up the dialog of figure 16.11. Starting in the `hello-world/` directory, or any

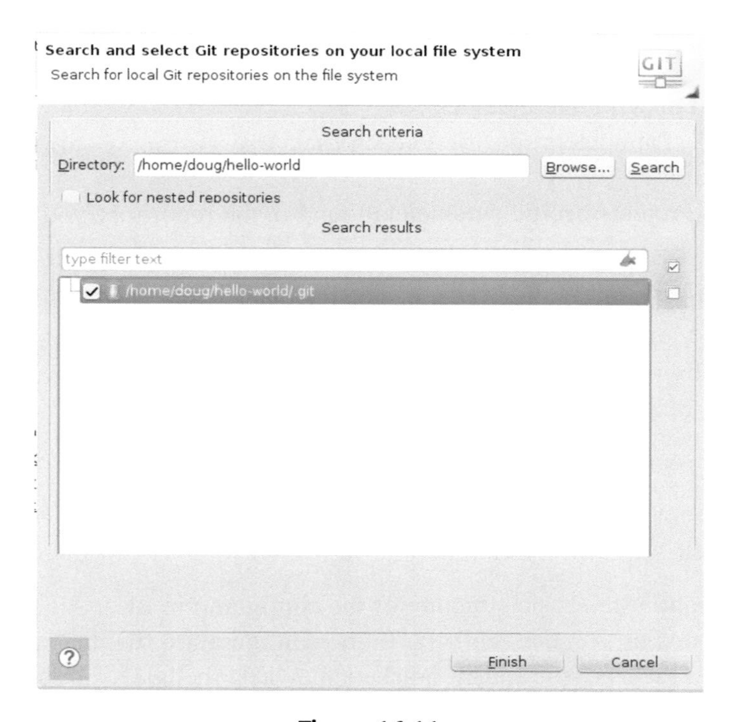


**Figure 16.11**
Add repository to Eclipse.

parent, click the Search button. Eclipse should find the `hello-world` repository. Select that entry and click OK. The repository will be added to the Git Repositories view.

Expand the `hello-world` entry and various sub-entries in the Git Repository view to reveal something like figure 16.12. The context menus for the objects shown here are pretty much what you would expect. If you right-click on `Branches` you can create a new branch. The currently checked out branch is identified by a checkmark. Right-click a different branch name to check it out.

In order to manage `hello-world` in the context of Eclipse, we need to make it an Eclipse project. Right-click the `Working Tree` entry and select `Copy Path to Clipboard`. Strictly speaking this isn't a C/C++ project, so we'll select `File>New>Project`, expand `General` and select `Project`. Click `Next` and give the project a name. "Hello World" comes to mind. Uncheck `Use default location`, right-click in the location box and select Paste. Then click `Finish`.

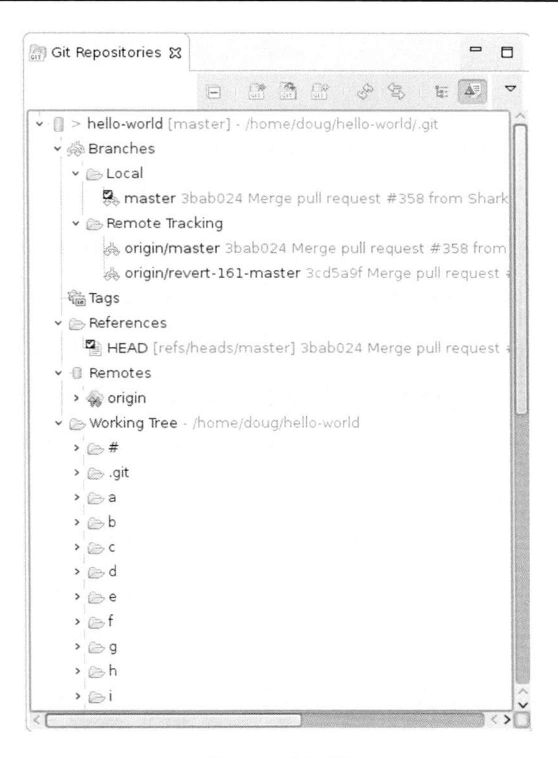


**Figure 16.12**
Git Repository view.

Make a change to one or more files. Just double-click the file entries under the Working directory to open them in the editor. Then right-click the changed entries to add them to the staging area. Now right-click the hello-world entry and select Commit. This brings up the dialog of figure 16.13. The Author and Committer fields are filled out for you and an editor is open to enter the commit message.

Modified files are listed below the Author and Committer entries. A check mark in status indicates that the file is staged. There's a check mark icon in the tool bar that selects all files. Note that since we created an Eclipse project, there's now a `.project` file in the working directory. Click Finish and the commit is saved.

### *Creating a new repository*

Ultimately, the point of all this is to bring your own projects under git control. As an example, let's create a new repository out of one of the Eclipse projects we were working with earlier in the book. How about the network project?

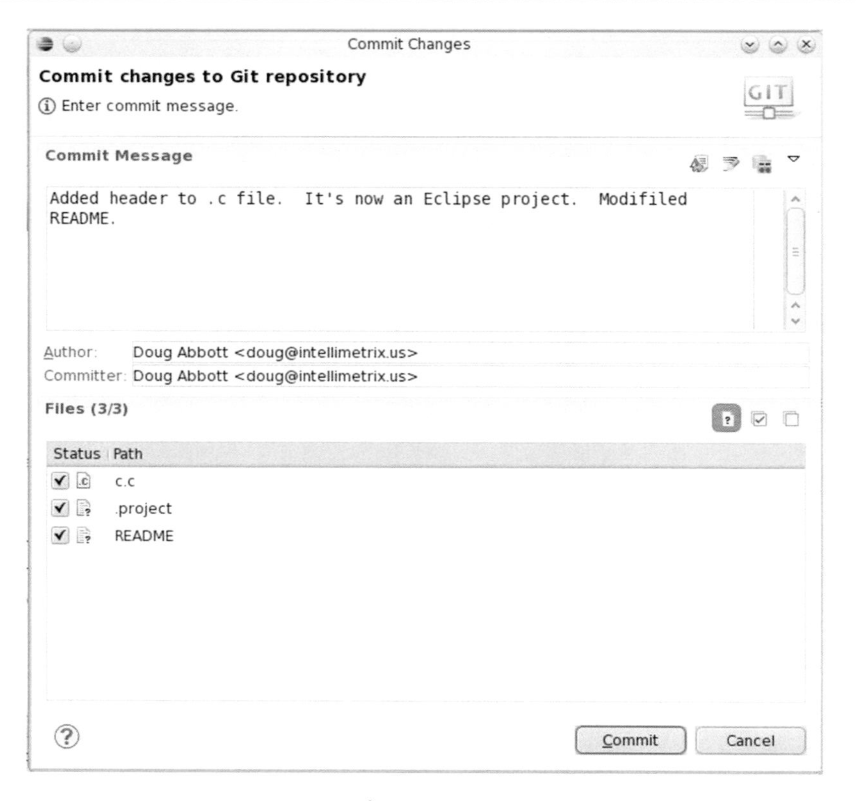


**Figure 16.13**
Commit dialog.

To get an existing Eclipse project into a git repository, select the project entry in Project Explorer, right-click and select `Team>Share Project`. This brings up the dialog of figure 16.14.

You'll probably want to create a new repository at this point. The question is, where? You can put the repository in the project's folder by checking Use or create repository in parent folder of project. Or you might want to put it in the parent of the project directories so that you can easily manage all of the projects in one git repository. When you click Create, you get a dialog that asks for a path and a name. The name ends up being a directory under path that must be empty or non-existent. That's where the `.git/` directory is stored. Your project is then moved into the repository directory when you click Finish.

Back in the Project Explorer view the network entry now has "[my_projects NO-HEAD]". This shows that the project is now managed by the my_projects repository but hasn't had an initial commit yet. If you right-click and select Team, one of the options is Commit,

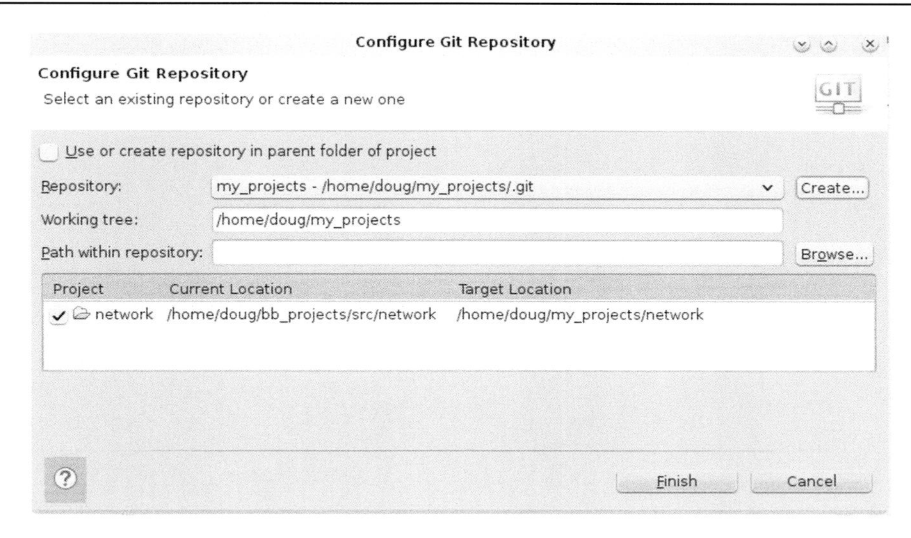


**Figure 16.14**
Configure git Repository dialog.

which brings up the Commit dialog that we saw in figure 16.13. Select all the files, enter a commit message, and click Commit.

You can still do most of your work from the C/C++ perspective. Git doesn't intrude very much on the development process. When you need to do "git like" stuff such as managing branches, then you need to switch to the Git perspective.

That wraps up our exploration of source code control using git. In the next and final chapter, we'll look at Linux and real-time.

## Resources

http://git-scm.com – The git home page

http://git-scm.com/book -- *Pro Git* by Scott Chacon. This is a very readable and complete introduction to git. The entire book is available in HTML on the git-scm site or you can download it as a PDF.

http://en.wikipedia.org/wiki/Comparison_of_revision_control_software -- A very nice comparison of about 30 version control systems including git.

https://github.com – A git hosting service that claims to be hosting over 3 million repositories from almost 2 million people. It's free to open source projects. Monthly subscription fees apply to private repositories.

http://git-scm.com/docs -- Among other thinsgs, you'll find a compact, handy little "cheat sheet" for using git as a PDF.

gitorious.org – Another git hosting service. This one seems to have more of a community feel than github.

CHAPTER 17

# Linux and real-time

## Chapter Outline



> *You can put racing stripes on a bulldozer but it won't go any faster.*
>
> ***Victor Yodaiken***

We briefly reviewed the concept of real-time programming in Chapter 1, The embedded and real-time space. Let me repeat the quote I included there:

> A real-time system is one in which the correctness of the computations not only depends upon the logical correctness of the computation but also upon the time at which the result is produced. If the timing constraints of the system are not met, system failure is said to have occurred.
>
> ***Donald Gillies in the Real-time Computing FAQ***

To put it another way, a late answer is a wrong answer. And remember, real-time does not necessarily mean real fast. It means *fast enough*, but more importantly, *reliably* fast enough. In practice, this means we can establish an upper bound on the time it takes the system to respond to an event. This is called *latency*.

## Hard versus Soft Real-Time

Real-time systems can be classified as either *hard* real-time or *soft* real-time. The primary difference between them is the consequences of missing a scheduling deadline. In hard

**Linux for Embedded and Real-time Applications.**
**DOI: http://dx.doi.org/10.1016/B978-0-12-811277-9.00017-1**

real-time, the scheduling deadline must be met every single time. Missing the deadline means the system has failed, possibly with catastrophic consequences.

A good example of hard real-time is a fly-by-wire flight control system where a computer intervenes between the pilot and the engine and control surfaces. The control algorithm depends on precisely timed samples of airspeed, altitude, rate of climb or descent, and so on. If these samples are late, the algorithm can become unstable and the plane crashes.

Soft real-time can tolerate missing a deadline occasionally, if an average latency is maintained. The scheduling deadline is more of a goal than an absolute. A good example of soft real-time is the network of automated teller machines. "Come on," I hear you say, "the ATM network isn't real-time." Oh yes it is, because when you put your card in that machine, you expect a response within a few seconds. So the designers of the ATM network have set a goal of responding in, say, 5 seconds. But they are not going to get terribly upset if occasionally it takes a few seconds longer. After all, what's the worst that can happen? You get annoyed.

## Why Linux is Not Real-Time

Linux was conceived and built as a general-purpose multiuser operating system in the model of Unix. The goals of a multiuser system are generally in conflict with the goals of real-time operation. General purpose operating systems are tuned to maximize average throughput even at the expense of latency, while real-time operating systems attempt to minimize, and place an upper bound on, latency, sometimes at the expense of average throughput.

There are several reasons why standard Linux is not considered suitable for real-time use:

- "Coarse-grained Synchronization": This is a fancy way of saying that kernel system calls are not preemptible. Once a process enters the kernel, it cannot be preempted until it is ready to exit the kernel. Actually, this is no longer the case. Much kernel code is now preemptible, and preemption is a kernel configuration option.
- Paging: The process of swapping pages in and out of virtual memory is, for all practical purposes, unbounded. We have no way of knowing how long it will take to get a page off of a disk drive, and so we simply cannot place an upper bound on the time a process may be delayed due to a page fault.
- "Fairness" in Scheduling: Reflecting its Unix heritage as a multiuser time-sharing system, the conventional Linux scheduler does its best to be fair to all processes. Thus, the scheduler may give the processor to a low priority process that has been waiting a long time, even though a higher priority process is ready to run. It should be noted, though, that the scheduler has improved substantially over the years to the point where standard Linux could be considered soft real-time.

- Request Reordering: Linux reorders I/O requests from multiple processes to make more efficient use of hardware. For example, hard disk block reads from a lower priority process may be given precedence over read requests from a higher priority process in order to minimize disk head movement or improve the chances of error recovery.
- "Batching": Linux will batch operations to make more efficient use of resources. For example, instead of freeing one page at a time when memory gets tight, Linux will run through the list of pages, clearing out as many as possible, delaying the execution of all processes.

Years ago the consequences of these issues in using Linux, or even Windows on a PC, were noticeable. Back in the day, moving the mouse while executing a compute-intensive function like rendering a complex graphics image would cause the mouse to occasionally stop and then jump, because the compute- or I/O-bound process had the CPU locked up. With today's high speed, multicore processors coupled with the vastly improved Linux scheduler, this is not nearly as noticeable.

The net effect of all these characteristics is that we cannot put an upper bound on the latency that a user task or process may encounter. By definition this is not real-time.

### *Measuring Latency: Cyclictest*

There is a tool called `cyclictest` that measures latency. The basic operation of cyclictest is relatively trivial, as shown in Listing 17.1. The program sleeps for some interval and when it wakes up, it gets the actual time and computes the difference, or latency, from what the time should be. It reports minimum, "actual," average, and maximum latency.

```
clock_gettime((&now))
next = now + interval

while (!shutdown)
{
        clock_nanosleep(&next):
        clock_gettime (&now);
        diff = calcdiff (now, next);

        # update stat-> min, max, total latency, cycles
        # update the histogram data
        next += interval;
}
```

**Listing 17.1**
Cyclictest algorithm.

The `cyclictest` binary is hard to find, so you will have to build it from sources. Follow these steps:

```
git clone git://git.kernel.org/pub/scm/linux/kernel/git/clrkwllms/rt-tests.git
cd rt-tests
make NUMA=0
su
./cyclictest -S
```

By default, `cyclictest` requires a package called `libnuma-dev`, where NUMA stands for Non-Uniform Memory Access. Most PCs do not implement NUMA, and so do not need the library otherwise. `NUMA=0` allows `cyclictest` to build without it.

The default behavior of `cyclictest` is to run one thread on each processor core at the lowest priority. Here is the result I get on my dual-core laptop where only one processor is dedicated to my Linux VM running CentOS 7 (kernel version 3.10.0-327.el7.x86_64:

```
# /dev/cpu_dma_latency set to 0us
policy: other/other: loadavg: 0.24 0.06 0.06 2/316 6204
T: 0 ( 6204) P: 0 I:1000 C: 10948 Min: 11 Act: 625 Avg: 304 Max: 18988
```

Latencies are reported in microseconds, so my system reported a minimum latency of 11 microseconds, average of 304, and a maximum of almost19 milliseconds.

Of course, the system is fairly lightly loaded right now. We can introduce an artificial load. Under `samples/src` is a directory `load50/`. Have a look at the file `load50.c`. This is a trivial program that forks off 50 processes that just spin. The number of processes forked is an argument to the program with a default value of 50.

Make `the load50` program, start `cyclictest` in one shell window, then run `load50` in another shell. You should see the maximum latency go up dramatically. On my system with a load of 20, the maximum went up to 3.5 seconds. I'm not sure I believe that. There may be a problem with the program.

### Improving Linux Latency

There are some things we can do to improve the latency of standard Linux. Specifically, we can change the kernel's "scheduling policy" and process priority for cyclictest, and we can lock the process's memory image into RAM so it won't be paged out.

The default scheduling policy, called `SCHED_OTHER`, uses a fairness algorithm and gives all processes using this policy priority 0, the lowest priority. This is fine for "normal" processes. The alternate scheduling policies are `SCHED_FIFO`, and `SCHED_RR`. These are intended for time-critical processes requiring lower latency. Processes using these alternate scheduling policies must have a priority greater than 0. Thus, a process scheduled with `SCHED_FIFO` or `SCHED_RR` will preempt any running normal process when it becomes ready.

A `SCHED_FIFO` process runs until it blocks or it yields. `SCHED_RR` is a minor variation on `SCHED_FIFO` that adds time slicing. If a `SCHED_RR` process exceeds its time slice it is placed at the back of its priority queue.

`cyclictest` has options to lock memory (`-m`) and to give its threads a higher priority (`-p` *N*). On my system, running `cyclictest` with a priority of 90 and a load of 20 kept the maximum latency down to about 12 milliseconds.

Locking memory did not improve things, which probably means there is no swapping going on anyway.

## Two Approaches

OK, so Linux is not real-time. What do we do about it? Well, there are at least two very different approaches to giving Linux deterministic behavior.

### Preemption Improvement

One approach is to modify the Linux kernel itself to make it more responsive. This primarily involves introducing additional preemption points in the kernel to reduce latency. An easy way to do this is to make use of the "spinlock" macros that already exist in the kernel to support symmetric multi-processing (SMP). In an SMP environment spinlocks prevent multiple processors from simultaneously executing a critical section of code on shared memory. In a uni-processor environment the spinlocks simply lock out the scheduler.

The preemption improvement strategy starts by turning the spinlocks into priority inheritance mutexes. This work of introducing preemption points in the kernel has been ongoing since late in the 2.4 series development. Mainline kernels these days implement three different preemption models, one of which is selected at build time as a configuration parameter:

- **No Forced Preemption (Server):** This is the traditional Linux preemption model, intended to maximize throughput. This is the preferred model for server applications. System call returns and interrupts are the only preemption points.
- **Voluntary Kernel Preemption (Desktop):** This option reduces the latency of the kernel by adding more "explicit preemption points" to the kernel code, at the cost of slightly lower throughput. In addition to explicit preemption points, system call returns and interrupt returns are still implicit preemption points.

- **Preemptible Kernel (Low-Latency Desktop):** This option reduces the latency of the kernel even further by making all kernel code that is not executing in a critical section preemptible. An implicit preemption point is located after each preemption disable section.

Note, by the way, that when we speak of kernel preemption, we are referring to *process* latency, not *interrupt* latency. With a standard uni-processor Linux kernel, interrupt latency is on the order of 10 microseconds or less, depending of course on processor speed. As noted above, maximum *process* latency for a standard kernel is in the tens of milliseconds. The preemption improvement strategy reduces that to one to two milliseconds.

The advantage to the preemption improvement approach is that the real-time applications run in user space, just like any Linux application using the familiar Linux/POSIX APIs. Real-time processes are subject to the same memory protection rules as ordinary processes. But it is still not hard real-time. Latency is reduced, but there are simply too many execution paths in the kernel to permit comprehensive analysis of determinism.

Along with kernel preemption, the process scheduler has undergone several major overhauls over the years. The original scheduler, in use through the 2.4 series, exhibited what is called O(n) behavior, meaning that the time required to schedule the next process was proportional to the number of ready processes in the queue. That is, the scheduler traversed the entire queue to determine which process had the highest claim on the processor at this instant in time.

So-called "big-O" notation is a way of representing the time complexity of algorithms as a function of the number of inputs. Many algorithms exhibit O(n) behavior, while others can be much worse, as in $O(n^2)$ or, really bad, O(n!). Most of the time, the best behavior is O (1), meaning that the execution time is constant with respect to the number of inputs. But, on the other hand, if the constant time is 3 hours, it might be better to look at an O(n) algorithm that may actually run faster for the most likely number of inputs.

The scheduler introduced in the 2.6 series exhibited O(1) behavior meaning that the scheduling time is independent of the number of ready processes. Kernel version 2.6.23 introduced the *Completely Fair Scheduler* (CFS), where the run queue is replaced by what is called a *Red-Black Tree*. It is an O(log n) algorithm. Choosing the next task to run is constant time, but inserting a preempted task back in the tree is log n. The CFS also features nanosecond timing granularity.

We should take a quick look at another configuration option that surfaced in the 2.6 kernel. You now have the option to turn off virtual memory. No more paging! This addresses the second major impediment to real-time behavior in Linux listed at the beginning of the chapter.

Turning off virtual memory is fine, as long as you can anticipate the maximum amount of memory your system will require. In fact, in an embedded situation this is often possible.

You know what applications will be running and how much memory they require. So just make sure that much RAM is available.

The option to turn off virtual memory is called "Support for paging of anonymous memory" and shows up under "General setup" in the configuration menu.

In parallel with all the work on preemption improvement and the scheduler, another project called the *PREEMPT_RT Patch* has been ongoing. The goal of this project is to implement true hard real-time behavior in the Linux kernel. The key point of the PREEMPT_RT patch is to minimize the amount of kernel code that is nonpreemptible, while also minimizing the amount of code that must be changed in order to provide this added preemptibility.

Another significant feature of the PREEMPT_RT patch is how it handles interrupts. The upper, A, portion of Fig. 17.1 shows how interrupts are normally handled. When the interrupt occurs, the interrupt handler of necessity preempts the currently running task to do whatever is necessary to service the interrupt. The interrupted task is held off for as long as it takes. In a PREEMPT_RT kernel, all the interrupt handler does is wake up a corresponding interrupt kernel thread, as shown in the B part of Fig. 17.1. The kernel thread is scheduled by the scheduler. It is possible that the currently running task has higher priority than the interrupt thread, so it continues to run following the interrupt handler. In principle, this gives us better control over scheduling.

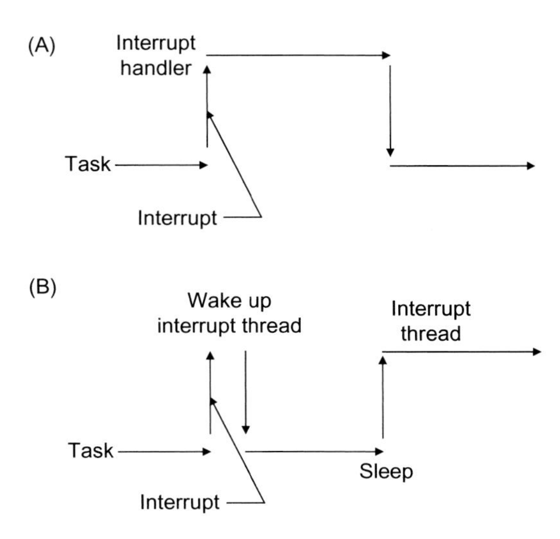


**Figure 17.1**
Interrupt handling with the PREEMPT_RT patch.

We will look at the PREEMPT_RT patch in more detail later in the chapter.

### Interrupt Abstraction

The improvements in the mainline kernel, while significant, still do not get us to deterministic, real-time performance. Even though the default scheduler is very good, it still attempts to be "fair." There is also the problem of the block I/O system reordering and combining I/O requests in the interest of maximizing throughput. So we need to circumvent these issues, at least for the parts of the application that are truly real-time.

It turns out that in a great many applications, only a small part of the system truly requires hard real-time determinism. Controlling a high-speed PID loop or moving a robot arm are examples of hard real-time requirements. But logging the temperature the PID loop is trying to maintain, or graphically displaying the current position of the robot arm, are generally not real-time requirements.

The alternate, and some would say more expedient, approach to real-time performance in Linux relies on this distinction between what is real-time and what is not. The technique is to run Linux as the lowest priority task (the idle task if you will) under a small real-time kernel. The real-time functions are handled by higher priority tasks running under this kernel. The non real-time stuff, like graphics, file management, and networking, which Linux already does very well, is handled by Linux.

This approach is called "Interrupt Abstraction," because the real-time kernel takes over interrupt handling from Linux. The Linux kernel "thinks" it is disabling interrupts, but it really is not. The essence of interrupt abstraction is illustrated in Fig. 17.2. The real-time kernel effectively intercepts hardware interrupts before they get to the Linux kernel. Linux no longer has direct control over enabling and disabling interrupts. So when Linux says disable interrupts, the RT kernel simply clears an internal software interrupt enable flag, but

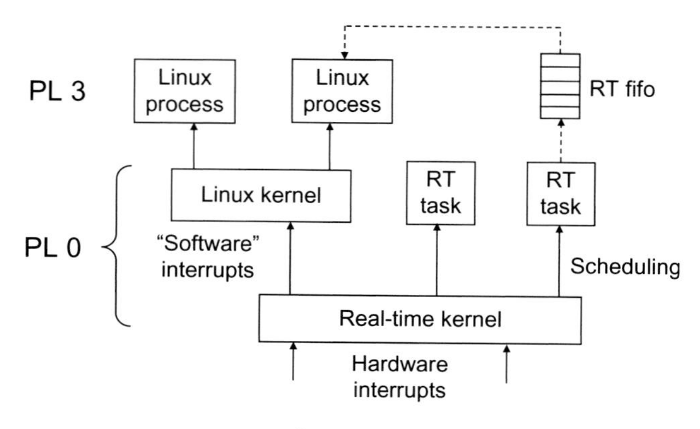


**Figure 17.2**
RT Linux architecture.

leaves interrupts enabled. When a hardware interrupt occurs, the RT kernel first determines to whom it is directed:

- RT Task: Schedule the task
- Linux: Check the software interrupt flag. If enabled, invoke the appropriate Linux interrupt handler. If disabled, note that the interrupt occurred and deliver it later when Linux re-enables interrupts.

The Linux kernel is treated as the lowest priority, or idle, task under the RT kernel, and only runs when there are no real-time tasks ready to run.

Of course, there will be times when the RT kernel has to disable hardware interrupts to manage its own critical sections, but these are of much shorter duration than the critical sections in Linux.

The real-time tasks can run either as kernel space loadable modules, or as user space processes. User space RT tasks communicate with the RT kernel through a kernel space "buddy task." Kernel space RT tasks will usually have some need to communicate with user-space processes for things like file access, network communication, or user interface. The RT kernel provides mechanisms like FIFOs and shared memory that support communication with user space processes.

Being much smaller and simpler, the real-time OS is amenable to execution time analysis that provides reliable upper bounds on latency. And while this approach also involves modifying the kernel, the extent of the modifications is substantially less than the Preemption Improvement approach.

The Interrupt Abstraction RTOS introduces its own API, and purists insist that this is not "true" Linux. But as the interrupt abstraction approach has evolved over the years, the two major implementations have both evolved "wrappers" around the native API for POSIX threads. This mitigates, to some extent, the objections of the purists.

There are two major implementations of the Interrupt Abstraction approach:

- RTLinux: This is the original interrupt abstraction implementation. It was developed at the New Mexico Institute of Mining and Technology under the direction of Victor Yodaiken. Even though RTLinux was developed as open source, Victor subsequently received a patent on the idea of running one operating system on top of another and formed a company, FSM Labs, to commercialize a proprietary version that he subsequently sold to Wind River. The open source continued to exist for several years, but its website now redirects to Wind River.
- RTAI: This is an enhancement of RT Linux developed at the Dipartimento di Ingeneria Aerospaziale, Politecnico di Milano under the direction of Prof. Paolo Mantegazza. It is a very active open source project with many contributors.

### A Third Way: Xenomai

Xenomai is another approach to interrupt abstraction that was, for a time, part of the RTAI project. It subsequently split off into an independent project. All real-time tasks in xenomai run in user space. Xenomai's real-time core implements a native API that is suspiciously similar to RTAI. In addition, it offers "skins" that implement APIs for popular OSes such as Posix, VxWorks, and μITRON, among others. What apparently drove the split between RTAI and xenomai is that the former seems obsessively focused on minimizing latency, while the latter aims for clean design, extensibility, and maintainability.

Xenomai actually implements two different approaches to interrupt abstraction based on a *Real Time Driver Model* (*RTDM*) that specifies characteristics for device drivers operating with real-time constraints. The traditional dual kernel approach is illustrated in Fig. 17.3. The real-time kernel is called the Cobalt Core, and is represented in user space by `libcobalt`. Applications using Posix threads APIs talk directly to `libcobalt`. Applications using other RTOS APIs talk to their respective "skin," which then talks to the copperplate interface.

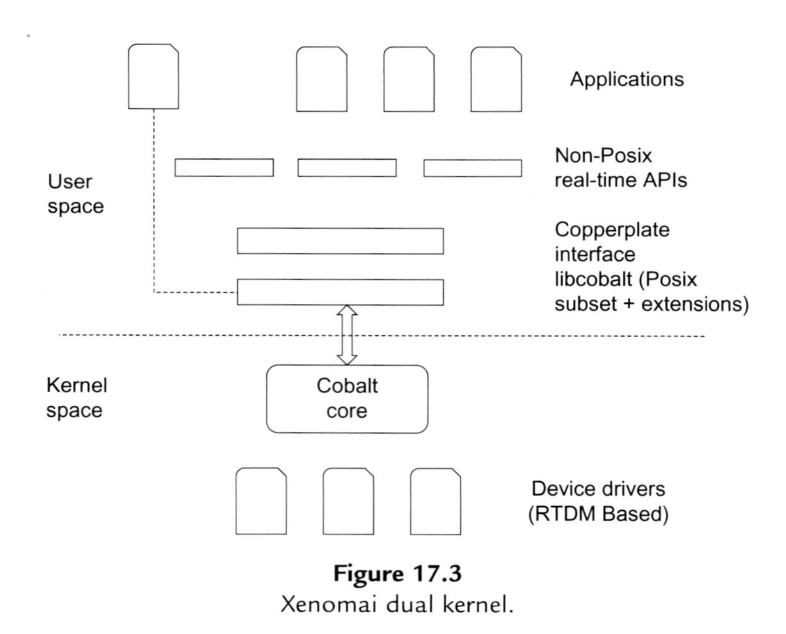


**Figure 17.3**
Xenomai dual kernel.

The alternate, newer approach dispenses with the second kernel and has `glibc` talk directly to the native RTDM specification (Fig. 17.4). All scheduling is done by the standard Linux kernel. The performance is not as good as the dual kernel approach, but it is perceived to be simpler. In most cases, you would probably want to install the PREEMPT_RT patch in your kernel. The choice of single or dual kernel operation is made at build time.

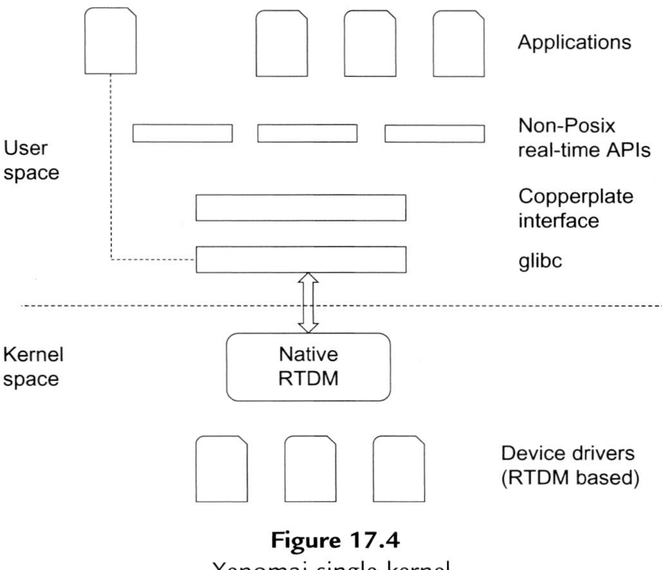


**Figure 17.4**
Xenomai single kernel.

## *Working With PREEMPT_RT*

The PREEMPT_RT patch is hosted at `kernel.org`. Navigate to `kernel.org//pub/linux/kernel/projects/rt/`. My intention was to patch and build the BeagleBoard kernel using an RT patch for the 3.8.13 kernel. Go to the `3.8/` subfolder. The patches are distributed in two forms:

`patch...` is all of the patches wrapped up in one big patch file
`patches...` is a collection of the individual patches

Both forms are available in either `.gz` or `.xz` format, where `.xz` is a more efficient algorithm. Oddly, the tar command in CentOS 7 does not recognize `.xz` as a valid compression format, but the KDE Ark program does, and will gladly decompress it.

I was skeptical of the patch on this page because it applies to kernel version 3.8.13.14, an extra level of versioning that we do not have. So I went into the `older/` subfolder and found a number of patches for the 3.8.13 kernel. Which one to pick? Well, I arbitrarily picked the `-rt9` patch, uncompressed it and applied it against the kernel we patched for the BeagleBone Black back in Chapter 12, Configuring and building the Linux kernel. Perhaps not surprisingly, some files did not patch correctly. Well, only 6 out of 365 files that were patched. Not bad actually.

I suspect that the failures were the result of the patches we did to the upstream 3.8.13 kernel for the BBB. Those confused `patch` when it tried to apply the PREEMPT_RT patch. When `patch` fails, it alerts you and also writes the parts that failed to `<file>.rej`. Then you can manually attempt to patch the file. Of the six files that failed, five were relatively straightforward to fix. The sixth, `drivers/net/ethernet/ti/cpsw.c`, had eight out of 45 “hunks” fail. This proved to be a “deal breaker.” I could only find where to apply some of the failed hunks. The kernel make command ran fine up to this file and then failed with a great many errors. Solving these errors will not be easy, and I’m not going to try to do it right now. If you need the PREEMPT_RT patch on your BBB, have at it.

The alternative then is to build the PREEMPT_RT patch into a workstation kernel. I happened to choose kernel version 3.18.48, but feel free to pick any version that strikes your fancy. The corresponding PREEMPT_RT patch applied cleanly.

Remember that it is always a good idea to have a known good `.config` file when building a new kernel. The configuration of your running desktop kernel is in the `/boot` directory. Copy that file to `.config` in your kernel source tree and run `make oldconfig`. Each new version of the kernel introduces new configuration options that are not yet in your `.config` file. `make oldconfig` steps you through all of the new options, asking you to select a value for each. There is a default, which is usually “no.” In going from 3.10 to 3.18 there were a *lot* of new options. I usually start out reading the new option descriptions, but after a while I just start hitting Enter until the command prompt comes back.

Run `make xconfig`. PREEMPT_RT introduces a few new options of its own. Under Processor type and features there are two new options for the Preemption Model radio button, as shown in Fig. 17.5. Select Fully Preemptible Kernel.

**Figure 17.5**
Preemption model options.

Save the configuration and execute these steps:

```
make
su
make modules_install
make install
```

The last step does several things:

1. Copies the kernel executable, vmlinuz, to /boot
2. Creates an initial RAM disk image and copies it to /boot
3. Creates an entry in the Grub configuration file for the new kernel

Reboot and select your newly built kernel at the Grub menu. Oddly, on a lightly loaded system (load50 not running) the maximum latency on my system went up to 34.5 milliseconds, while the average went down to 284 microseconds. Starting load50 20 and then running cyclictest with −m and −p 90 yielded `Min: 1 Avg: 348 Max: 37086`. Better than the loaded results on the stock kernel, but still not great. I'm not sure I believe the minimum of 1. There must be something more subtle going on.

## Wrap Up

We've come to the end of our long journey through the wonderful world of embedded Linux. I hope you have found it worthwhile, maybe even fun. The adventure started with some introductory material on the nature of embedded and real-time, and an introduction to Linux. After installing Linux we configured our workstation to support a cross-development environment for developing applications for an ARM-based single board computer.

Next, we explored Eclipse, the open source IDE and found, among other things, that it has a very slick graphical front end for GDB, the Gnu debugger. Then we set up our target board, the BeagleBone Black, and launched into application programming for said board, getting into Posix threads, networking, and graphics programming using QT.

Then we looked at some of the components and tools that make up the embedded Linux ecosystem. We configured and built a Linux kernel. We reviewed the emerging trend of integrated build environments for embedded Linux that pretty much attempt to do everything for you. We looked at BusyBox, the "Swiss army knife of Linux," and saw how Linux initializes with the goal of getting our application to start at boot time. We explored git, the source code control tool that has become pretty much a standard in the open source world. And finally, in this chapter, we evaluated why Linux is not considered real-time, and what we can do about that.

By now you have probably come to realize that embedded Linux is a *big* topic. My objective with this book is to introduce you to at least some of the variety of issues you will encounter in getting into embedded Linux. We haven't pursued any of them in great detail. I encourage you to explore those topics that catch your interest more thoroughly.

You may have noticed that I have been fairly frank about the difficulties I encountered in developing the examples for this book. Some authors would be reluctant to admit to that, feeling that it would somehow diminish their credibility. I did it to make the point that this Linux stuff is *not easy*! In fact, it is downright difficult. After something like 15 years of mucking around with Linux, I still spend a fair amount of time on Google teasing out error messages or other bizarre behavior that initially makes no sense. Get used to it, guys. You are going to bump up against a lot of walls. In the end it is worth it. Linux really is a great tool for developing embedded devices.

But remember that it is not necessarily the only tool. Back in Chapter 1, The embedded and real-time space, I suggested that there are applications that do not require the extensive resources of Linux. If your application falls into that category, then by all means take a look at the alternatives offered there.

Whatever your role in the world of embedded systems, have fun with it. In the end, that is what it is all about. Despite my love-hate relationship with Linux, I'm still having fun with it, because I'm constantly learning. So have fun and make some great products.

## Resources

rt.wiki.kernel.org/index.php/Cyclictest – Learn more about cyclictest.
wiki.linuxfoundation.org/realtime/start – The Linux Foundation is now the community resource for the PREEMPT_RT patch.
www.rtai.org – Home page for the Real-Time Application Interface.
www.rtlinuxfree.com. – This used to be the site for the open source version of RTLinux. It now redirects to Wind River.
www.xenomai.org – Home page for the Xenomai project.
www.gna.org/adeos – Adeos is another project that came out of RTAI. It is an interrupt pipeline, or "nanokernel," that allows several operating systems to run in parallel. It is the lowest level in the xenomai Cobalt Core.

# Appendix A: U-boot Commands

Das u-boot supports an extensive command set. This section describes the most useful of those commands. Because u-boot is highly configurable, not all commands are necessarily in any given implementation. The behavior of some commands is also configuration-dependent, and may depend on environment variable values.

All u-boot commands expect numbers to be entered in hexadecimal, with or without the leading 0x. In the following descriptions, arguments are bracketed by <>. Optional arguments are further bracketed by []. Command names can generally be abbreviated to the shortest unique string, shown in parentheses.

## Information Commands

`bdinfo (bdi)`—Displays information about the target board such as memory sizes and locations, clock frequencies, MAC address, etc. This information is available to the Linux kernel.

`coninfo (conin)`—Displays information about the available console I/O devices. It shows the device name, flags, and the current usage.

`iminfo <start addr> (imi)`—Displays header information for images such as Linux kernels or ramdisks. It lists the image name, type, and size, and verifies the CRC32 checksum. `iminfo` takes an argument that is the starting address of the image. The behavior of `ininfo` is influenced by the `verify` environment variable.

`help [<command name>]`—Self-explanatory. Without an argument it prints a short description of all commands. With a command name argument it provides more detailed help.

## Memory Commands

By default, the memory commands operate on 32-bit integers. They can be made to operate on 16-bit words or 8-bit bytes by appending ".w" (for 16 bits) or ".b" (for 8 bits) to the command name. There is also a ".l" suffix to explicitly specify 32 bits. Example: `cmp.w 1000 2000 20` compares 32 (20 hex) words at addresses 0x1000 and 0x2000.

`base [<offset>] (ba)`—Get or set a base address to be used as an offset for all other memory commands. With no argument it displays the current base, default is 0. A numeric argument becomes the new base. This is useful for repeated accesses to a specific section of memory.

`crc32 <start addr><length> [<store addr>] (crc)`—Computes and displays a 32-bit checksum over `<length>` bytes, starting at `<start addr>`. The optional third argument specifies a location to store the checksum.

`cmp <addr1><addr2><count>`—Compares two regions of memory. `<count>` is the number of *data items* independent of the specified size, byte, word, or long.

`cp <source><dest><count>`—Copies a range of memory to another location. `<count>` is the number of items copied.

`md <addr> [<count>]`—Memory display. Displays a range of memory in both hex and ASCII. The optional `<count>` argument defaults to 64 items. `md` remembers the most recent `<addr>` and `<count>`, so that if it is entered without an address argument, it continues from where it left off.

`mm <start addr>`—Modify memory with autoincrement. Displays the address and current contents of an item, and prompts for user input. A valid hexadecimal value replaces the current value and the command displays the next item. This continues as long as you enter valid hex values. To terminate the command, enter a non-hex value.

`mtest <start addr><end addr> [<pattern>]`—Simple read/write RAM test. This test modifies RAM and may crash the system if it touches areas used by u-boot, such as the stack or heap.

`mw <addr><value> [<count>]`—Writes `<value>` to a range of memory starting at `<addr>`. The optional `<count>` defaults to 1.

`nm <addr>`—Modify memory at a constant address (no autoincrement). Interactively writes valid hex data to the same address. This is useful for accessing I/O device registers. Terminate the command with a non-hex input.

`loop <addr><count>`—Reads memory in a tight loop. This can be useful for scoping. This command **never terminates!** The only way out is to reset the board.

## NAND Flash Memory Commands

NAND flash is organized into arbitrary sized named *partitions*. All NAND commands are prefixed by the keyword `nand`. The BeagleBone Black u-boot implements the NAND commands, even though there is no NAND device installed.

`nand info`— Show available NAND devices.

`mtdparts`—List NAND partitions.

`nand erase [clean] [<offset><size>]`—Erase `<size>` bytes starting at `<offset>`. If `<offset>` not specified, erase entire chip.

`nand scrub`—*Really* erase the entire chip, including bad blocks. Considered "unsafe."

`nand createbbt`—Create bad block table.

`nand bad`—List bad blocks.

`nand read[.raw] <address><offset> | <partition><size>` – Read `<size>` bytes from flash, either from numerical `<offset>` or `<partition>` name into RAM at `<address>`. The `.raw` modifier bypasses ECC to access the flash as is.

`nand write[.raw] <address><offset> | <partition><size>` – Write `<size>` bytes from `<address>` in RAM to flash either at numerical `<offset>` or `<partition>` name. The `.raw` modifier bypasses ECC to access the flash as is.

## Execution Control Commands

`boot`—Boot default, that is `run bootcmd`.

`bootd` – Boot default, that is `run bootcmd`.

`bootm <addr> [<param>...]`—Boots an image such as an operating system from memory. The image header, starting at `<addr>`, contains the necessary information about the operating system type, compression, if any, load, and entry point addresses. `<param>` is one or more optional parameters passed to the OS. The OS image is copied into RAM and uncompressed if necessary. Then control is transferred to the entry point address. For Linux, two optional parameters are recognized: the address of an initrd RAM disk image, and the image of a flattened device tree (FDT) blob. To boot a kernel with a FDT but no initrd, pass the second argument as "-".

Note, incidentally, that images can be booted from RAM, having been downloaded, for example, with TFTP. In this case, be careful that the compressed image does not overlap the memory area used by the uncompressed image.

`bootz <addr> [initrd[:size]] [fdt]`—Boots a Linux zImage stored in memory at `<addr>`. `<initrd>` is the address of an initrd image of `<size>` bytes. `<fdt>` is the address of a flattened device tree blob. To boot a kernel with a FDT but no initrd, pass the second argument as "-".

go <addr> [<param>...]—Transfers control to a "stand-alone" application starting at <addr> and passing the optional parameters. These are programs that do not require the complex environment of an operating system.

## Download Commands

Three commands are available to boot images over the network using TFTP. Two of them also obtain an IP address before executing the file download.

bootp <load_addr> [<filename>]—Obtain an IP address using the bootp protocol, then download <filename> to <load_addr>. If <filename> is not given, the value of bootfile will be used.

nfs <load_addr> [<filename>] [[<hostIPaddr>:]<filename>]—Boot an image over the network using the NFS protocol. If <hostIPaddr> is not given, the value of serverip is used. If <filename> is not given, the value of bootfile will be used.

tftpboot <load_addr><filename> (tftp)—Just download the file. Assumes client already has an IP address, either statically assigned or obtained through DHCP.

dhcp—Get an IP address using DHCP.

It is also possible to download files using a serial line. The recommended terminal emulator for serial image download is kermit, as some users have reported problems using minicom for image download.

loadb <offset>—Accept a binary image download to address <offset> over the serial port. Start this command in u-boot, then initiate the transmission on the host side.

loads <offset>—Accept an S-record file download to address <offset> over the serial port.

## Environment Variable Commands

printenv [<name>...]—Prints the value of one or more environment variables. With no argument, printenv lists all environment variables. Otherwise, it lists the value(s) of the specified variable(s).

setenv <name> [<value>]—With one argument, setenv removes the specified variable from u-boot's environment and reclaims the storage. With two arguments it sets variable <name> to <value>. These changes take place in RAM only. **Warning:** use a space between <name> and <value>, not " = ". The latter will be interpreted literally with rather strange results.

Standard shell quoting rules apply when a value contains characters with special meaning to the command line parser, such as "$" for variable substitution and ";" for command separation. These characters are "escaped" with a back slash, "\". Example:

```
setenv netboot tftp 21000000 uImage\; bootm
```

`saveenv`—Writes the environment to persistent storage.

`run <name> [...]`—Treats the value of environment variable `<name>` as one or more u-boot commands and executes them. If more than one variable is specified, they are executed in order.

`bootd (boot)`—A synonym for run bootcmd to execute the default boot command.

## Environment Variables

The u-boot environment is kept in persistent storage and copied to RAM when u-boot starts. It stores environment variables used to configure the system. The environment is protected by a CRC32 checksum. This section lists some of the environment variables that u-boot recognizes.

The variables shown here serve specific purposes in the context of u-boot and, for the most part, will not be used explicitly. When needed, u-boot environment variables are used explicitly in commands, much the same way that they are in shell scripts and makefiles, by enclosing them in $().

`autoload`—If set to "no", or any string beginning with "n", the `rarp` and `bootp` commands will only get configuration information and not try to download an image using TFTP.

`autostart`—If set to "yes", an image loaded using the `rarpb`, `bootp`, or `tftp` commands will be automatically started by internally calling the `bootm` command.

`baudrate`—A *decimal* number that specifies the bit rate for the console serial port. Only a predefined list of baudrate settings is available. Following the `setenv baudrate <n>` command, u-boot expects to receive a newline character at the new rate before actually committing the new rate. If this fails, the board must be reset and reverts to the old baud rate.

`bootargs`—The value of this variable is passed to the Linux kernel as boot arguments, i.e., the "command line."

`bootcmd`—Defines a command string that is automatically executed when the initial countdown is *not* interrupted, but only if the variable `bootdelay` is also defined with a non-negative value.

bootdelay—Wait this many seconds before executing the contents of the bootcmd variable. The delay can be interrupted by pressing any key before the bootcmd sequence starts. **Warning:** setting bootdelay to 0 executes bootcmd immediately, and effectively disables any interaction with u-boot. On the other hand, setting this variable to −1 disables auto boot.

bootfile—Name of the default image to be loaded by any of the download commands.

ethaddr—MAC address for the first or only Ethernet interface on the board, known to Linux as eth0. A MAC address is 48 bits, represented as six pairs of hex digits separated by dots.

eth1addr, eth2addr—MAC addresses for the second and third Ethernet interfaces when present.

ipaddr—Set an IP address on the target board for TFTP downloading.

loadaddr—Default buffer address in RAM for commands like tftpboot, loads, and bootm. Note: it appears, but does not seem to be documented, that there is a default load address, 0x82000000, built into the code.

serverip—IP address of the TFTP or NFS server used by the tftpboot and nfs commands.

serial#—A string containing hardware identification such as type, serial number, etc. This variable can only be set once, often during manufacturing of the board. U-boot refuses to delete or overwrite this variable once it has been set.

verify—If set to "n" or "no", disables the checksum calculation over the complete image in the bootm command to trade speed for safety in the boot process. The header checksum is still verified.

The following environment variables can be automatically updated by the network boot commands (bootp, dhcp, or tftp) depending on the information provided by your boot server:

bootfile—See above
dnsip—IP address of your Domain Name Server
gatewayip—IP address of the Gateway (Router)
hostname—Target host name
ipaddr—See above
netmask—Subnet mask
rootpath—Path to the root filesystem on the NFS server
serverip—See above
filesize—Size in bytes, as a hex string, of the file downloaded using the last bootp, nfs, or tftp command.

# Appendix B: Why Software Should Not Have Owners

Digital information technology contributes to the world by making it easier to copy and modify information. Computers promise to make this easier for all of us.

Not everyone wants it to be easier. The system of copyright gives software programs "owners," most of whom aim to withhold software's potential benefit from the rest of the public. They would like to be the only ones who can copy and modify the software that we use.

The copyright system grew up with printing—a technology for mass production copying. Copyright fitted in well with this technology, because it restricted only the mass producers of copies. It did not take freedom away from readers of books. An ordinary reader, who did not own a printing press, could copy books only with pen and ink, and few readers were sued for that.

Digital technology is more flexible than the printing press: when information has a digital form, you can easily copy it to share it with others. This very flexibility makes a bad fit with a system like copyright. That is the reason for the increasingly nasty and draconian measures now used to enforce software copyright. Consider these four practices of the Software Publishers Association (SPA):

- Massive propaganda saying it is wrong to disobey the owners to help your friend.
- Solicitation for stool pigeons to inform on their coworkers and colleagues.
- Raids (with police help) on offices and schools, in which people are told they must prove they are innocent of illegal copying.
- Prosecution (by the US government, at the SPA's request) of people such as MIT's David LaMacchia, not for copying software (he is not accused of copying any), but merely for leaving copying facilities unguarded and failing to censor their use.

All four practices resemble those used in the former Soviet Union, where every copying machine had a guard to prevent forbidden copying, and where individuals had to copy information secretly and pass it from hand to hand as "samizdat." There is, of course, a difference: the motive for information control in the Soviet Union was political; in the United States the motive is profit. But it is the actions that affect us, not the motive. Any attempt to block the sharing of information, no matter why, leads to the same methods and the same harshness.

Owners make several kinds of arguments for giving them the power to control how we use information:

- Name calling.

  Owners use smear words such as "piracy" and "theft," as well as expert terminology such as "intellectual property" and "damage," to suggest a certain line of thinking to the public—a simplistic analogy between programs and physical objects.

  Our ideas and intuitions about property for material objects are about whether it is right to *take an object away* from someone else. They do not directly apply to *making a copy* of something. But the owners ask us to apply them anyway.
- Exaggeration.

  Owners say that they suffer "harm" or "economic loss" when users copy programs themselves. But the copying has no direct effect on the owner, and it harms no one. The owner can lose only if the person who made the copy would otherwise have paid for one from the owner.

  A little thought shows that most such people would not have bought copies. Yet the owners compute their "losses" as if each and every one would have bought a copy. That is exaggeration—to put it kindly.
- The law.

  Owners often describe the current state of the law, and the harsh penalties they can threaten us with. Implicit in this approach is the suggestion that today's law reflects an unquestionable view of morality—yet at the same time, we are urged to regard these penalties as facts of nature that cannot be blamed on anyone.

  This line of persuasion is not designed to stand up to critical thinking; it is intended to reinforce a habitual mental pathway.

  It is elementary that laws do not decide right and wrong. Every American should know that, 40 years ago, it was against the law in many states for a black person to sit in the front of a bus; but only racists would say sitting there was wrong.
- Natural rights.

  Authors often claim a special connection with programs they have written, and go on to assert that, as a result, their desires and interests concerning the program simply outweigh those of anyone else—or even those of the whole rest of the world. (Typically companies, not authors, hold the copyrights on software, but we are expected to ignore this discrepancy.)

To those who propose this as an ethical axiom—the author is more important than you—I can only say that I, a notable software author myself, call it bunk.

But people in general are only likely to feel any sympathy with the natural rights claims for two reasons.

One reason is an overstretched analogy with material objects. When I cook spaghetti, I do object if someone else eats it, because then I cannot eat it. His action hurts me exactly as much as it benefits him; only one of us can eat the spaghetti, so the question is, which? The smallest distinction between us is enough to tip the ethical balance.

But whether you run or change a program I wrote affects you directly and me only indirectly. Whether you give a copy to your friend affects you and your friend much more than it affects me. I shouldn't have the power to tell you not to do these things. No one should.

The second reason is that people have been told that natural rights for authors is the accepted and unquestioned tradition of our society.

As a matter of history, the opposite is true. The idea of natural rights of authors was proposed and decisively rejected when the US Constitution was drawn up. That is why the Constitution only *permits* a system of copyright and does not *require* one; i.e., why it says that copyright must be temporary. It also states that the purpose of copyright is to promote progress—not to reward authors. Copyright does reward authors somewhat, and publishers more, but that is intended as a means of modifying their behavior.

The real established tradition of our society is that copyright cuts into the natural rights of the public—and that this can only be justified for the public's sake.

- Economics.

The final argument made for having owners of software is that this leads to production of more software.

Unlike the others, this argument at least takes a legitimate approach to the subject. It is based on a valid goal—satisfying the users of software. And it is empirically clear that people will produce more of something if they are well paid for doing so.

But the economic argument has a flaw: it is based on the assumption that the difference is only a matter of how much money we have to pay. It assumes that "production of software" is what we want, whether the software has owners or not.

People readily accept this assumption because it accords with our experiences with material objects. Consider a sandwich, for instance. You might well be able to get an equivalent sandwich either free or for a price. If so, the amount you pay is the only difference. Whether or not you have to buy it, the sandwich has the same taste, the same nutritional value, and in either case you can only eat it once. Whether you get the sandwich from an owner or not cannot directly affect anything but the amount of money you have afterwards.

This is true for any kind of material object—whether or not it has an owner does not directly affect what it *is*, or what you can do with it if you acquire it.

> But if a program has an owner, this very much affects what it is, and what you can do with a copy if you buy one. The difference is not just a matter of money. The system of owners of software encourages software owners to produce something—but not what society really needs. And it causes intangible ethical pollution that affects us all.

What does society need? It needs information that is truly available to its citizens—e.g., programs that people can read, fix, adapt, and improve, not just operate. But what software owners typically deliver is a black box that we cannot study or change.

Society also needs freedom. When a program has an owner, the users lose freedom to control part of their own lives.

And above all, society needs to encourage the spirit of voluntary cooperation in its citizens. When software owners tell us that helping our neighbors in a natural way is "piracy," they pollute our society's civic spirit.

This is why we say that free software is a matter of freedom, not price.

The economic argument for owners is erroneous, but the economic issue is real. Some people write useful software for the pleasure of writing it or for admiration and love; but if we want more software than those people write, we need to raise funds.

For 10 years now, free software developers have tried various methods of finding funds, with some success. There is no need to make anyone rich; the median US family income, around $35k, proves to be enough incentive for many jobs that are less satisfying than programming.

For years, until a fellowship made it unnecessary, I made a living from custom enhancements of the free software I had written. Each enhancement was added to the standard released version, and thus eventually became available to the general public. Clients paid me so that I would work on the enhancements they wanted, rather than on the features I would otherwise have considered highest priority.

The Free Software Foundation (FSF), a tax-exempt charity for free software development, raises funds by selling GNU CD-ROMs, T-shirts, manuals, and deluxe distributions (all of which users are free to copy and change), as well as from donations. It now has a staff of five programmers, plus three employees who handle mail orders.

Some free software developers make money by selling support services. Cygnus Support, with around 50 employees (when this article was written), estimates that about 15% of its staff activity is free software development—a respectable percentage for a software company.

Companies including Intel, Motorola, Texas Instruments, and Analog Devices have combined to fund the continued development of the free GNU compiler for the language C.

Meanwhile, the GNU compiler for the Ada language is being funded by the US Air Force, which believes this is the most cost-effective way to get a high quality compiler. (Air Force funding ended some time ago; the GNU Ada Compiler is now in service, and its maintenance is funded commercially.)

All these examples are small; the free software movement is still small, and still young. But the example of listener-supported radio in this country (the United States) shows it is possible to support a large activity without forcing each user to pay.

As a computer user today, you may find yourself using a proprietary program. If your friend asks to make a copy, it would be wrong to refuse. Cooperation is more important than copyright. But underground, closet cooperation does not make for a good society. A person should aspire to live an upright life openly with pride, and this means saying "No" to proprietary software.

You deserve to be able to cooperate openly and freely with other people who use software. You deserve to be able to learn how the software works, and to teach your students with it. You deserve to be able to hire your favorite programmer to fix it when it breaks.

You deserve free software.



Updated: $Date: 2001/09/15 20:14:02 $ $Author: fsl $

# Index

*Note*: Page numbers followed by "*f*" and "*t*" refer to figures and tables, respectively.

## S



## T



## U



## V



## W



## X



## Y



## Z

Printed in the United States
By Bookmasters